Mammalogy Techniques Lab Manual

Mammalogy Techniques Lab Manual

James M. Ryan

JOHNS HOPKINS UNIVERSITY PRESS

Baltimore

Unless otherwise indicated, all drawings and photographs were produced by the author.

Johns Hopkins University Press
2715 North Charles Street
Baltimore, Maryland 21218-4363
www.press.jhu.edu

Library of Congress Cataloging-in-Publication Data

Names: Ryan, James M. (James Michael), 1957- author.
Title: Mammalogy techniques lab manual / James M. Ryan.
Description: Baltimore : Johns Hopkins University Press, 2018. |
   Includes bibliographical references and index.
Identifiers: LCCN 2017054171 | ISBN 9781421426075 (paperback : alk. paper) |
   ISBN 1421426072 (paperback : alk. paper) | ISBN 9781421426082 (electronic) |
   ISBN 1421426080 (electronic)
Subjects: LCSH: Mammalogy—Laboratory manuals. | Mammalogy—Technique.
Classification: LCC QL703.R93 2018 | DDC 599.078—dc23
LC record available at https://lccn.loc.gov/2017054171

A catalog record for this book is available from the British Library.

*Special discounts are available for bulk purchases of this book. For more information, please
contact Special Sales at 410-516-6936 or specialsales@press.jhu.edu.*

Johns Hopkins University Press uses environmentally friendly book materials, including
recycled text paper that is composed of at least 30 percent post-consumer waste,
whenever possible.

Contents

Acknowledgments

This manual comes out of exercises I've taught over many years to students at Hobart and William Smith Colleges. I am extremely grateful to the many students who worked through these exercises with thoughtful suggestions and good cheer. I am also thankful for the community of students, faculty, and researchers that make up the American Society of Mammalogists. A kinder, more generous group of colleagues would be very hard to find.

I am grateful to the following individuals who provided data or editorial suggestions for this manual. Dr. Rick Mace generously provided the raw GPS data for tracking a female grizzly bear in Chapter 14. A subset of that data is used in the exercises in this manual. Dr. Roland Kays provided valuable suggestions and editorial comments on several chapters. I am very grateful to the Wisconsin State Laboratory of Hygiene for generously providing the human karyotype photos and to Russ Crutcher of Microlab Northwest for the hair photos. Thanks to both Dr. Robert Huber for permission to use his Java DataGrinders Applet and to the Sevilleta National Wildlife Refuge in New Mexico for access to the rodent parasite data set. Many thanks go to Adam Schneider for his excellent web tool GPSVisualizer and to the many researchers who help develop and maintain MoveBank. For nursing this project forward, I am grateful to Vincent J. Burke, Tiffany Gasbarrini, Juliana McCarthy, and Lauren Straley at Johns Hopkins University Press and to copy editor Heidi Fritschel.

Finally, I also offer my sincerest gratitude to my wife, Gillian, for her patience and support.

Mammalogy Techniques Lab Manual

1 Introduction

Background

Mammals inhabit nearly every continent and every sea. Some are tiny and others are massive. They have adapted to life underground, in the frozen Arctic, in the hottest deserts, and in every habitat in between. Some remain active year-round, while others spend part of the year in hibernation. Mammals are terrestrial, arboreal, fossorial, aquatic, and, in the case of bats, aerial. In sum, mammals are a diverse and fascinating group.

Most mammalogists (those who study mammals) were drawn to the group through field work and a love of the outdoors. Not surprisingly, they want to share those experiences with their students. However, field work is becoming difficult at many institutions owing to large class size, distance to available field sites, and liability issues. Nevertheless, it is important for instructors to engage students early and often in field experiences.

Mammalogists are keenly aware of the need to get students back into the field. In 2007 mammalogist Mark Hafner published a paper lamenting the decline in field-based studies of natural history and mammalogy over the past few decades. He noted, "We can no longer assume that our students, many of whom are nature-deprived, have a set of basic outdoor skills that will safeguard them in the field; these students need to be taught about field-work from the ground up" (Hafner, 2007). The purpose of this manual is to provide opportunities to practice some of the techniques used by today's mammalogists. An additional goal is to get students outdoors, where they can hone their observation skills and develop the essential tools to become the next generation of practicing mammalogists.

Most chapters in this manual provide field exercises. When it is not feasible to collect data in the field, this manual also provides exercises that are readily adapted to laboratory or classroom learning. There are, however, no labs on identifying the key features of the major mammal groups. Most textbooks cover those details, and it would be impossible to include such information for all regions in North America (or the world). In addition, some groups are defined largely based on molecular characters, making it difficult to study in a 3- to 4-hour laboratory exercise. The assumption is that instructors who wish to include identification of mammals will do so on their own.

How to Use This Manual

This manual is broken into chapters emphasizing particular principles or skills. Each chapter begins with a short set of learning objectives and a list of equipment needed for the exercises. These are followed by an explanation of the concept covered. The background information is necessarily brief because lectures typically cover much of this material and because the goal is to get students working on the concepts as quickly as possible. The background information should be sufficient to provide students with a working knowledge of the content of each of the exercises. Terms are generally explained when first introduced, but students are encouraged to refer to the glossary when necessary.

Following the chapter introduction, a set of exercises is presented for each concept or skill. Most chapters include both outdoor and indoor exercises to accommodate more urban institutions, unpredictable weather, and other logistical issues. When possible, the outdoor activities should be conducted first, followed by one or more of the indoor exercises. In some cases, when students have access to computers for data analysis, the exercises can be assigned as homework. In addition, several large data sets are included on the website that accompanies this text (jhupbooks.press.jhu.edu/content/mammalogy-techniques-lab-manual) for those instances when students are not able to gather sufficient data in the field. For example, in Chapter 14 a large set of GPS tracking data for a female grizzly bear (*Ursus arctos*) is available for download so that students may complete the lab without collecting tracking data of their own.

Several exercises use statistical models that may be unfamiliar to students. Within these exercises, equations are given and an example is worked through. The objective is to give all students, regardless of math experience, a concrete example to help demystify the concept.

In some cases, only calculators are required. In a few cases the exercise helps students build an Excel spreadsheet that they can use to solve the problem described in the exercise (and that can be used again if they choose to do honors or postgraduate research).

Where software is introduced to ease analysis of large or complex data sets, open-source (free) or web-based tools are presented. The primary reason is to minimize cost to students. A secondary reason is to ensure that students have access to these software tools on any platform (Windows, Mac, Linux) after they graduate and no longer have access to institutional resources. Several of the exercises rely on the programing language R.

R Statistical Environment

R is a language for statistics, data analysis, and graphics. It is open source, meaning that it is free to use, and it is available for free in source code form under the terms of the Free Software Foundation's GNU General Public License. More important, R is highly extensible: users can develop new tools that add to the R environment. It runs on a wide variety of platforms including Linux, Windows, and Mac OS.

If you are looking for a robust, comprehensive statistical and data analysis program, I recommend that you learn R. R is rapidly becoming the dominant data analysis tool for scientists of all kinds. Not only is it used by Google, Facebook, and other major companies, it is increasingly used by government agencies like the US Fish and Wildlife Service, the US Geological Survey, and others. This means that the community of users is wide and growing. Learning R now will pay dividends later.

Installation files and documentation for Windows, Mac, and Linux can be found at the Comprehensive R Archive Network at http://cran.r-project.org/. Download R from this site first. At first, some features of R may seem a bit odd. To make it somewhat easier to use, I recommend using RStudio, a graphical user interface for R. RStudio combines a syntax editor, a graphic window, and other features to make using R more intuitive. RStudio also has a free open-source edition for desktop use. It can be downloaded at www.rstudio.com/products/rstudio/download/.

Like any major data analysis tool (SPSS, SAS), R has a learning curve. This manual assumes that instructors have some familiarity with R. However, step-by-step instructions are given in most exercises, and R scripts are provided in the resources that accompany this text. One of the most useful features in R is the use of R packages. Packages are assembled chunks of code, data, and documentation created and shared by users of R, and thousands of packages are available for free. Packages are installed in R when you need them and make data analysis much easier. Most of the exercises in this manual make use of R packages. Another important feature of R is its ability to import data from many formats. You can import data from text files (.txt), Excel files (.xls or .csv), SPSS, SAS, MySQL databases, and many other formats. These features represent just the tip of the iceberg in terms of what R can do. For beginners, the following list of books may be useful:

- Braun, W., and D. Murdoch. (2007) *A First Course in Statistical Programming with R*. Cambridge, UK: Cambridge University Press.
- Dalgaard, P. (2008) *Introductory Statistics with R*. 2nd ed. New York: Springer.
- Faraway, J. J. (2005) *Linear Models with R*. Boca Raton, FL: Chapman & Hall/CRC Press.
- Gardener, M. (2012) *Statistics for Ecologists Using R and Excel: Data Collection, Exploration, Analysis, and Presentation*. Exeter, UK: Pelagic Publishing.
- Gardener, M. (2014) *Community Ecology: Analytical Methods Using R and Excel*. Exeter, UK: Pelagic Publishing.
- Muenchen, R. A. (2009) *R for SAS and SPSS Users*. Springer Series in Statistics and Computing. New York: Springer.
- Murrell, P. (2005) *R Graphics*. Boca Raton, FL: Chapman & Hall/CRC Press.
- Stevens, M. H. (2009) *A Primer of Ecology with R*. New York: Springer.

There are also a number of useful websites for those wanting to learn R. They include

- DataCamp Introduction to R course: https://www.datacamp.com/courses/free-introduction-to-r/?tap_a=5644-dce66f&tap_s=10907-287229
- Quick-R: http://www.statmethods.net/index.html
- R-bloggers: https://www.r-bloggers.com/how-to-learn-r-2/
- The R Project for Statistical Computing: https://www.r-project.org
- The R Journal: https://journal.r-project.org/index.html

2 Mammal Skulls

Time Required
- One 3- to 4-hour lab period

Learning Objectives
- Learn the anatomy of the mammalian skull.
- Use the anatomical terms in a taxonomic key to identify mammalian skulls.
- Understand the structure of a dichotomous key.
- Use the traits in the skull key to identify unknown specimens.

Equipment Required
- A collection of mammal skulls from a variety of mammalian orders.
- A collection of bolts, nuts, and screws.
- Calipers or other measurement tools.

Background

There are more than 6,400 living species of mammals grouped into 29 orders. Because many species (especially species of small mammals) appear similar to the untrained eye, scientists have developed dichotomous keys to facilitate identification of species. Dichotomous keys are pairs of statements that allow the user to easily identify an unknown organism. The keys are arranged in "couplets" consisting of two statements that ask the user to make a choice about the presence, absence, or aspect of a particular characteristic of their unknown organism. That choice leads the user to a new couplet in the key, and the process repeats until a couplet choice leads to the name of the organism (or the group to which it belongs). Each couplet ends either in a number indicating which couplet to go to next in the key or in the name of the taxa. It is good practice to record your path through the key, including the final identification of your taxa (e.g., 1a; 2b; 3b; 4a; African elephant), so that you may retrace your steps if you make a mistake or if the key is long and complex.

Most dichotomous keys use scientific terminology to describe features. Thus, it is often important to familiarize yourself with basic anatomical terminology before attempting to use a key. Mammal skulls are often used in keys to identify species. However, mammal skulls are complex structures made up of numerous bones, processes, holes, and teeth (usually). Each feature is important and has been shaped by natural selection. For example, the diverse feeding habits of mammals are reflected in their teeth and jaw morphology. Teeth may be specialized for grasping, crushing, or grinding; the position and structure of the teeth often provide valuable clues to the animal's diet (see Chapter 3).

Bones and Features of the Skull

The major bones of the mammalian skull are listed below and illustrated in Figures 2.1, 2.2, and 2.3.

Alisphenoid	Pterygoid
Orbitosphenoid	Lacrimal
Dentary bone of the mandible	Basisphenoid
Palatine	Squamosal (temporal)
Ethmoid	Maxilla
Parietal	Tympanic bullae (composite)
Frontal	Nasal
Premaxilla	Vomer
Interparietal	Occipital
Presphenoid	Zygomatic arch (composite)
Jugal	

The following list of terms provides a basic introduction; your instructor may supplement this list with additional terms.

- **Crest**, a narrow prominent ridge.
- **Condyle**, a smooth, rounded projection for articulation with another bone.
- **Foramen**, a hole in a bone (typically, it passes a nerve or blood vessels).
- **Fossa**, a shallow depression or trench on the surface of a bone.
- **Labial**, on the side adjacent to the lips.
- **Lingual**, on the side adjacent to the tongue.

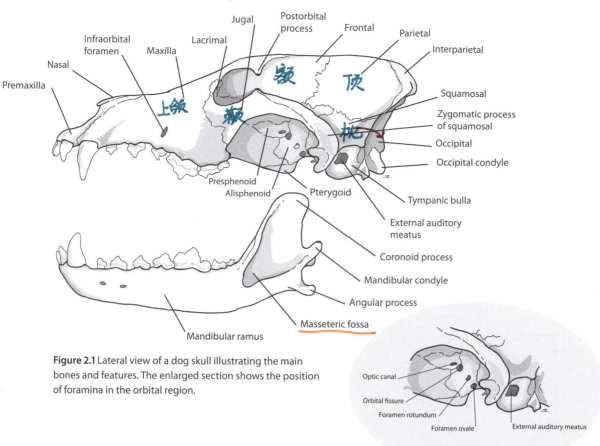

Nasal

Premaxilla

Infraorbital foramen

Maxilla

Lacrimal

Jugal

Postorbital process

Frontal

Parietal

Interparietal

顴 額 顶
上領 顳 旭

Squamosal

Zygomatic process of squamosal

Occipital

Occipital condyle

Presphenoid

Alisphenoid

Pterygoid

Tympanic bulla

External auditory meatus

Coronoid process

Mandibular condyle

Angular process

Masseteric fossa

Mandibular ramus

Figure 2.1 Lateral view of a dog skull illustrating the main bones and features. The enlarged section shows the position of foramina in the orbital region.

Optic canal

Orbital fissure

Foramen rotundum

Foramen ovale

External auditory meatus

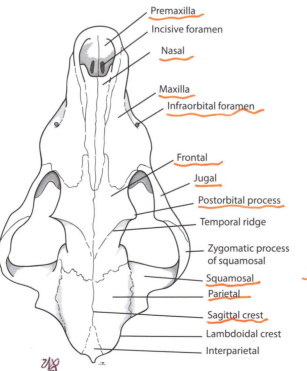

Premaxilla

Incisive foramen

Nasal

Maxilla

Infraorbital foramen

Frontal

Jugal

Postorbital process

Temporal ridge

Zygomatic process of squamosal

Squamosal

Parietal

Sagittal crest

Lambdoidal crest

Interparietal

脊

Figure 2.2 Dorsal view of a dog skull illustrating the main bones and features.

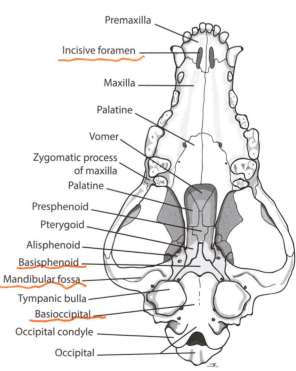

Premaxilla

Incisive foramen

Maxilla

Palatine

Vomer

Zygomatic process of maxilla

Palatine

Presphenoid

Pterygoid

Alisphenoid

Basisphenoid

Mandibular fossa

Tympanic bulla

Basioccipital

Occipital condyle

Occipital

Figure 2.3 Ventral view of a dog skull illustrating the main bones and features.

腹

- **Line**, a narrow raised ridge.
- **Meatus**, a small tubular opening.
- **Process**, a small projection or bump.
- **Septum**, a bony fence that separates two regions.
- **Sulcus**, a groove.
- **Suture**, the line formed by the junction of two bones.
- **Symphysis**, the cartilaginous junction or articulation formed between two bones; this junction or articulation may fuse (e.g., the two dentary bones fuse to form the mandibles at the mandibular symphysis).
- **Trochanter**, a large rounded projection for muscle attachment.

Variation in Mammalian Skulls

Mammal skulls vary tremendously in size and shape, and it is important to recognize the position of each bone regardless of species. Fortunately, the bones retain the same relative positions. For example, compare the canid skulls in Figures 2.1 through 2.3 with those of the pronghorn (*Antilocapra americana*) in Figures 2.4 to 2.6. Notice that the nasal bones still form the anterior margin of the nasal opening and are bordered by the premaxilla and maxilla (laterally) and the frontals (posteriorly). The palatine bone is still posterior to the maxilla on the roof of the mouth (palate). Also, notice that the pronghorn has unusually large orbits, projecting away from the skull. Their large eyes give them a wide field of view for detecting predators. Artiodactyls tend to have long preorbital skulls (rostrums) formed by elongated nasals and frontals. The frontal bones typically bear horn cores (sometimes with the horn sheath still attached) or antler peduncles (in adult males of the family Cervidae). A postorbital bar is usually also present, serving to fully enclose the orbit. Another feature common to many artiodactyls is the loss of upper incisors and canines (retained in some artiodactyls).

Zygomatic Morphology in Rodents

Rodents also have a great deal of variation in their skull morphology. A typical rodent skull (*Peromyscus*) is shown in Figure 2.7. Rodents (order Rodentia) are characterized by a pair of continuously growing upper and lower incisors. The incisors are used for gnawing and are beveled to a chisel-shape at the tips. The constant wear on the incisors is necessary because these teeth are ever growing. If, for example, one incisor in an upper pair is broken and no longer able to occlude with the lower incisor, then the opposing lower incisor will continue to grow. In some circumstances this incisor grows so far as to prevent the mouth from closing, and the animal starves to death. Rodent incisors are separated from the cheek teeth by a long diastema. The nasals are long and narrow and, in many species, extend anterior to the incisors.

The upper and lower incisors come together during gnawing; their sharp, beveled edges allow the incisors to chisel away hard materials. Such gnawing requires powerful jaw muscles that act to move the lower jaw forward, bringing the lower incisors into occlusion with the uppers. Rodents have evolved unique arrangements of jaw muscles to accomplish this task. Primitively, the masseter muscle was divided into superficial, lateral, and

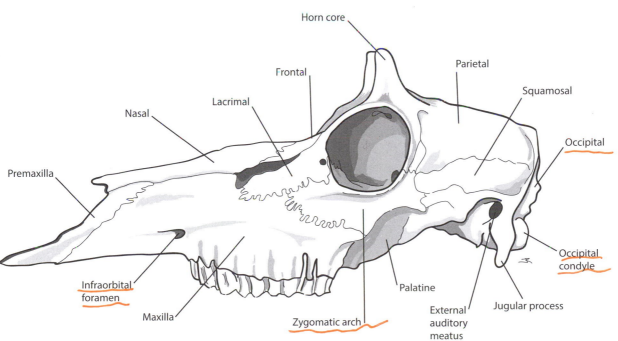

Figure 2.4 Lateral view of the skull of a pronghorn (Artiodactyla, *Antilocapra americana*).

Horn core

Frontal

Parietal

Lacrimal

Squamosal

Nasal

Occipital

Premaxilla

Occipital condyle

Infraorbital foramen

Palatine

Jugular process

Maxilla

Zygomatic arch

External auditory meatus

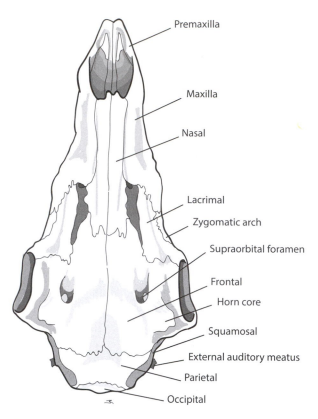

Premaxilla

Maxilla

Nasal

Lacrimal

Zygomatic arch

Supraorbital foramen

Frontal

Horn core

Squamosal

External auditory meatus

Parietal

Occipital

Figure 2.5 Dorsal view of the skull of a pronghorn (*Antilocapra americana*).

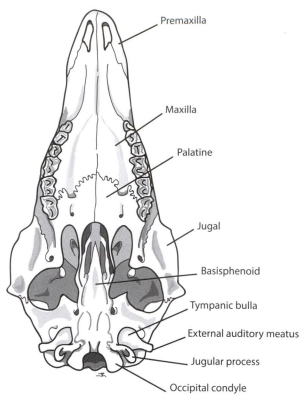

Premaxilla

Maxilla

Palatine

Jugal

Basisphenoid

Tympanic bulla

External auditory meatus

Jugular process

Occipital condyle

Figure 2.6 Ventral view of the skull of a pronghorn (*Antilocapra americana*).

Figure 2.7 Dorsal, ventral, and lateral views of the skull of a *Peromyscus* (Rodentia) showing the beveled incisors and diastema.

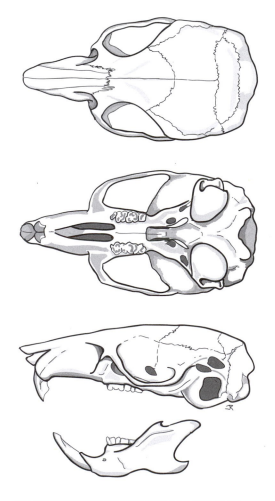

medial masseters, with the superficial masseter originating near the anterior end of the upper tooth row, the lateral masseter originating from the middle of the zygomatic arch, and the smaller medial masseter originating along the inside of the zygomatic arch (Figure 2.8). In this primitive configuration, the incisors could be used for only relatively simple gnawing movements. Rodents with this ancestral condition are called protrogomorphous. Protrogomorphy was common among early rodents, but only the mountain beaver (or sewell), *Aplodontia rufa*, retains this arrangement among living rodents.

As rodents continued to rely on gnawing, selection favored a more anterior position for at least some slips of the masseter. By moving a part of the masseter onto the rostrum in front of the zygomatic arch, rodents gained mechanical advantage during gnawing. These changes set the stage for the rapid radiation of rodents into the many species we recognize today. However, rodents accomplished this anterior repositioning of the jaw muscles in at least three different ways.

Sciuromorphous rodents (Figure 2.9) have the lateral masseter originating from a broad region at the front of the zygomatic arch (termed the zygomatic plate). The medial masseter is large and extends posteriorly along the zygomatic arch (relative to that in protrogomorphous masseters). This condition is found in beavers, squirrels, gophers, kangaroo rats and their allies (family Heteromyidae), and several other rodent families.

Infraorbital foramen

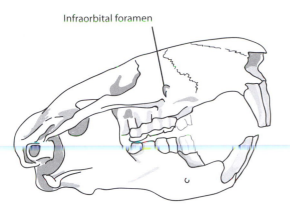

Superficial masseter

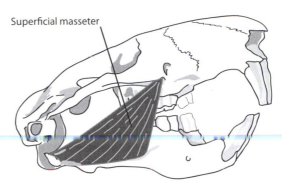

Figure 2.8
The masseter 咬肌
arrangement in
<mark>protrogomor-
phous</mark> (ancestral)
rodents, showing
the relative size of
the infraorbital
foramen and the
positions of the
masseter slips.

Lateral masseter

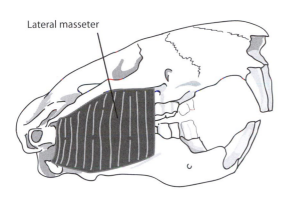

Medial masseter

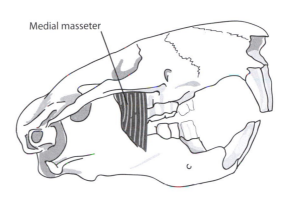

Zygomatic plate   Infraorbital foramen

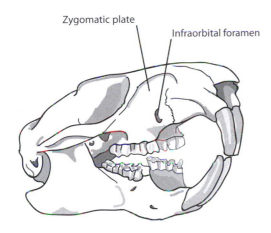

Superficial masseter

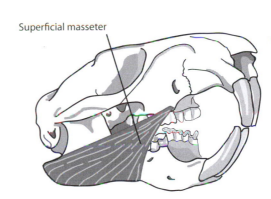

Figure 2.9
The masseter
arrangement in
<mark>sciuromorphous</mark>
rodents, showing
the relative size of
the infraorbital
foramen and the
positions of the
masseter slips.

Lateral masseter

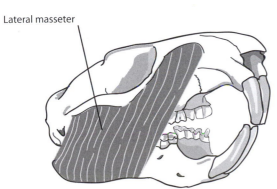

Medial masseter

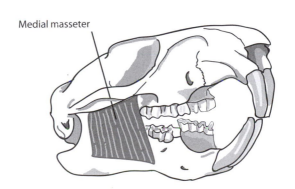

Figure 2.10 The masseter arrange- ment in hystrico- morphous (*top*) and myomor- phous (*bottom*) rodents.

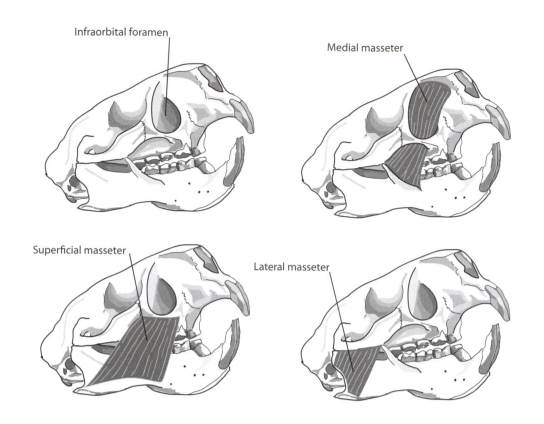

Infraorbital foramen

Medial masseter

Superficial masseter

Lateral masseter

Hystricomorphous masseter

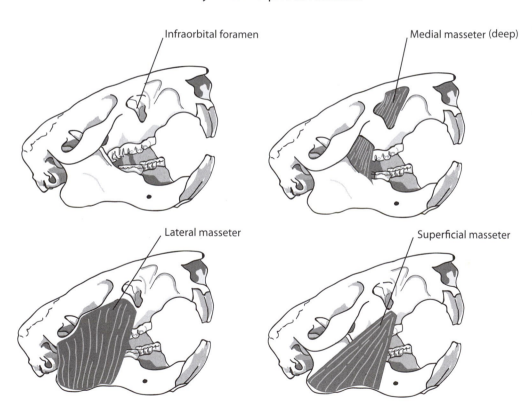

Infraorbital foramen

Medial masseter (deep)

Lateral masseter

Superficial masseter

Myomorphous masseter

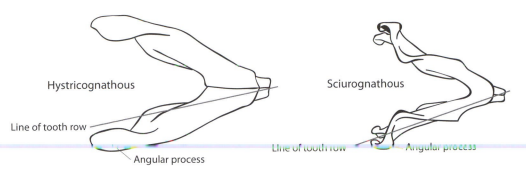

Hystricognathous

Line of tooth row

Angular process

A

Sciurognathous

Line of tooth row   Angular process

B

Figure 2.11
A comparison of the position of the angular process of the lower jaw in (**A**) a hystricognathous rodent and (**B**) a sciurognathous rodent.

The infraorbital foramen is greatly enlarged in the hystricomorphous masseter arrangement (Figure 2.10). The greatly expanded portion of the medial masseter originates on the lateral wall of the upper rostrum and passes ventrally and posteriorly through the greatly enlarged infraorbital foramen to insert on the mandible. Hystricomorphous masseters occur in New and Old World porcupines, guinea pigs, and many other rodent families.

A third myomorphous condition also involves a slip of the masseter passing through an enlarged infraorbital foramen (Figure 2.10). In this case, however, there is a zygomatic plate below a modestly enlarged infraorbital foramen. It is worth noting that it is still not certain how each masseter arrangement evolved, but it is clear that rodents independently evolved several solutions to the problem of repositioning the masseters for gnawing.

Today, mammalogists rely less on masseter morphology for classifying rodents and more on the morphology of the angular process of the lower jaw (e.g., hystricognathy versus sciurognathy) and on other morphological and molecular characters. In the hystricognathous condition, the angular process of the dentary is lateral to the plane of the tooth row, whereas in the sciurognathous jaw the angular process is approximately in a line with the rest of the tooth row (Figure 2.11).

Telescoping in Cetaceans

Another unusual skull morphology evolved in cetaceans in response to a fully aquatic lifestyle. Cetaceans may have highly telescoped and/or asymmetrical skulls. Toothed whales (Odontocetes) have telescoped skulls (Figure 2.12), where the external nares are repositioned on the top of the skull. This allows these cetaceans to breathe with just the nostrils (blow holes) exposed above the water. Telescoping involves the elongation of the rostrum and the posterior displacement of the nasals, maxillae, and premaxillae (among others) relative to the braincase. Odontocetes also have asymmetrical skulls. The bones on the right side of the skull are larger than their counterparts on the left side.

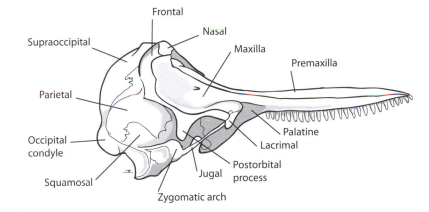

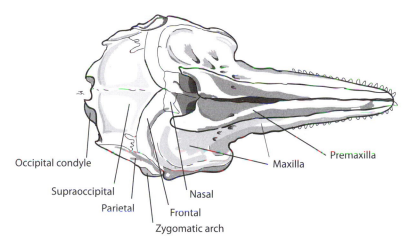

Frontal

Nasal

Supraoccipital

Maxilla

Premaxilla

Parietal

Palatine

Lacrimal

Occipital condyle

Postorbital process

Jugal

Squamosal

Zygomatic arch

Occipital condyle

Premaxilla

Maxilla

Supraoccipital

Nasal

Parietal

Frontal

Zygomatic arch

Figure 2.12 The asymmetrical and telescoped skull of a bottlenose dolphin (*Tursiops truncatus*). 错开的

不对称的

Skull Measurements

Mammalogists use a variety of standard skull measurements to aid in species identification, determine sex, establish age or maturity, and make geographic comparisons (Figures 2.13 to 2.15). Measurements are typically taken with dial calipers (for increased accuracy). Important standard measurements are the following:

- Greatest length of skull (greatest skull length, total skull length, maximum length of skull): from the

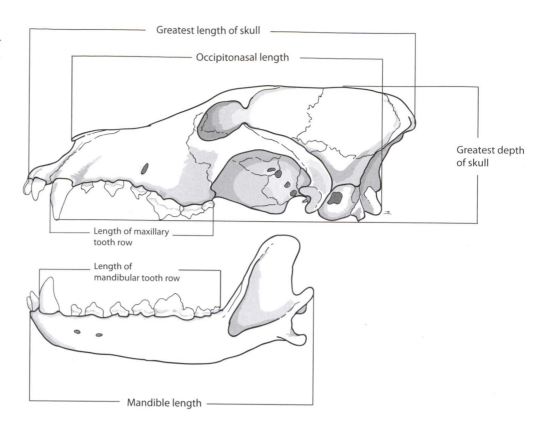

Figure 2.13 Skull measurements for a dog skull shown in lateral view.

Greatest length of skull

Occipitonasal length

Greatest depth of skull

Length of maxillary tooth row

Length of mandibular tooth row

Mandible length

most anterior part of the rostrum (excluding teeth) to the most posterior point of the skull.

- Greatest depth of skull: greatest length from the ventral-most margin of the skull (including teeth) to the dorsal-most margin of the skull.
- Occipitonasal length: distance from the anterior-most tip of nasal bone to the posterior-most margin of the occipital bone.
- Length of maxillary tooth row: distance from the anterior margin of the upper canine tooth to the posterior-most margin of the last molar on the maxilla.
- Length of mandibular tooth row: distance from the anterior margin of the lower canine tooth to the posterior-most margin of the last molar on the mandible.
- Mandible length: total length of the mandible from the alveolus of the incisors to the posterior-most margin of the mandible.
- Basal length: from the anterior-most margin of the premaxilla to the anterior-most margin of the foramen magnum.

- Breadth of braincase: greatest width across the braincase posterior to the zygomatic arches.
- Condylobasal length: from a line connecting the posterior-most projections of the occipital condyles to the anterior edge of the premaxillae.
- Incisive foramina length: greatest length of the anterior palatal (incisive) foramina.
- Palatal length: from the anterior edge of premaxilla to the anterior-most point on the posterior edge of palate.
- Postpalatal length: from the anterior-most margin of the posterior edge of the palate to the anterior-most margin of the foramen magnum.
- Zygomatic breadth: greatest distance between the outer margins of the zygomatic arches.
- Interorbital breadth: least distance across the top of the skull anterior to the postorbital process/bar.
- Postorbital breadth: greatest distance across the top of the skull at the level of the postorbital process/bar.

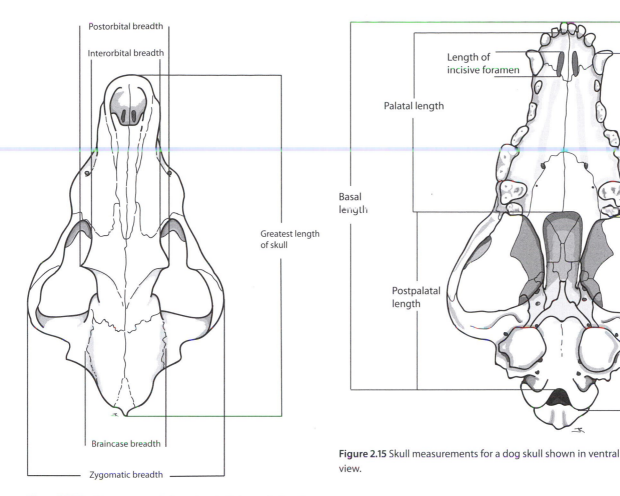

Figure 2.14 Skull measurements for a dog skull shown in dorsal view.

Figure 2.15 Skull measurements for a dog skull shown in ventral view.

Exercise 1: The Nuts and Bolts

As an example of how to construct and use a dichotomous key, you will begin with a set of simple "organisms" like those shown in Figure 2.16. These "organisms" are various sizes and shapes of nails, screws, and bolts (those provided by your instructor may differ).

To begin organizing this set of "organisms" into a key, first divide the set into two groups based on some easily identifiable feature. Record the two choices in the format of a dichotomous key, and repeat the process until all of the "organisms" have been identified (Table 2.1).

Your instructor will provide you with a set of hypothetical organisms similar to those in the key above. Working individually or with a partner, create a dichotomous key based on the characteristics of various organisms. Use easy-to-identify features and describe them in your key. When you are finished, your key will be given to another group, and they will use it to attempt to identify the organisms.

Figure 2.16 A set of "organisms" used in the example key.

1. Obtain a group of "organisms."
2. Assign each organism a unique name or letter.
3. Carefully examine or think about the characteristics of each organism.

Table 2.1 A sample dichotomous key for "organisms" similar to those shown in Figure 2.16

#	Character	Go to
1A	Shaft is straight	2
1B	Shaft is threaded	3
2A	Head is flat	Common nail
2B	Head is not flat	Finishing nail
3A	Head is dome-shaped	4
3B	Head is not dome-shaped	5
4A	Shaft is more than 10 times the width of head	Long bolt
4B	Shaft is less than 10 times the width of head	Short bolt
5A	Head is cone-shaped (not hexagonal)	6
5B	Head is hexagonal	7
6A	Shaft is entirely threaded	Wood screw
6B	Shaft is not threaded near head	Deck screw
7A	Shaft is pointed at tip	8
7B	Shaft is blunt at tip	Carriage bolt
8A	Shaft is short and head is relatively thick	Small lag screw
8B	Shaft is long and head is relatively thin	Long lag screw

4. Divide the group of organisms into two piles based on the presence or absence of a trait, and construct the first couplet in the key based on that trait.

5. Divide each group of organisms from step 4 into two new groups, and for each new group create a new couplet based on the presence or absence of a trait. Continue subdividing the groups and creating couplets until each organism has been named. Refer to Table 2.1 as an example. Use only one page for your key.

6. Make a second page as a blank key. In each spot on the right side where you have the final identity of an organism, leave a blank for the other groups to fill in (e.g., do not list the name of the organism).

7. When you are finished, you will be given the key created by another group. Try to identify each organism, and fill in the blanks in their key by following their key precisely.

8. Assign the other group's key a score: A score of 1 is very poor and a score of 5 is excellent. Provide an explanation for your score by answering the following questions:

- Where are the difficulties in the way traits were described in the key?
- How might those difficult areas be reworded to improve the meaning?
- If you were given a new "organism" that you had not used in the key (but similar), could the key be easily modified to add the new "species"?

Exercise 2: Dichotomous Keys of Skulls

Now put theory into practice. You will be given a series of mammal skulls by your instructor. Look them over carefully. Working individually or with a partner, create a dichotomous key of these skulls by using the skull features and measurements described earlier in this chapter. Use easy-to-identify features and describe characters in your key, but be sure to use appropriate terminology or measurement descriptions. For example, if you want to divide them into two groups of skulls based on size, you would measure the greatest length of skull for each skull and make a determination as to where to cut the length to define the two groups. You might say:

- 2A Greatest length of skull is greater than 4 cm.
- 2B Greatest length of skull is less than 4 cm.

Steps in constructing the key:

1. Obtain a group of skulls.
2. Assign each skull a unique letter.
3. Carefully examine or think about the characteristics of each skull.
4. Divide the group of skulls into two piles based on the presence or absence of a trait, and construct the first couplet in the key based on that trait.
5. Repeat step 4 until all of the skulls have been entered into the key.
6. Use the blank worksheet provided in Table 2.2 in the appendix to this chapter.

Exercise 3: Mystery Mammal Skull

Your instructor will provide you with several "mystery" skulls for you to identify. You will also be given a copy of *A Key to the Skulls of North American Mammals* by Bryann Glass and Monte Thies (1997) to help you in keying out your mystery skulls. The key is divided into two parts; the first part is a key to mammalian orders (pages 12-13). Once you know the order, you will turn to the specific key to that order. For example, if you have keyed out a skull as belonging to the order Carnivora, then you would turn to the key beginning on page 29 and continue. (Note that order Insectivora is no longer used in modern mammalian classifications; see Feldhamer et al., 2015, or Vaughan et al., 2015, for details.) Be sure to keep a record of the path you follow through the key. This is important; without a path, it is difficult to retrace your steps when resolving misidentifications. Use the glossary and figures provided here and those in Glass and Thies (1997). If terms or descriptions remain confusing, ask your instructor.

Appendix

Table 2.2 Dichotomous key worksheet

Skull ID	
Path followed	
Name of mammal	

Skull ID	
Path followed	
Name of mammal	

Skull ID	
Path followed	
Name of mammal	

Couplet	Description	Go to

3   Mammalian Teeth

Time Required
- One 3- to 4-hour lab period

Learning Objectives
- Learn the basic evolution of mammalian teeth.
- Recognize cusp patterns on mammalian teeth.
- Link diet to tooth morphology in mammalian species.
- Identify the different types of teeth: incisors, canines, premolars, and molars.
- Understand the diversity in mammalian cheek teeth.
- Write dental formulas for mammalian species.

Equipment Required
- A collection of mammal skulls from a variety of orders

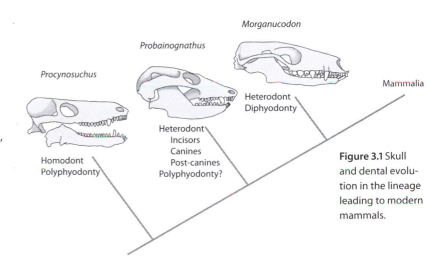

Figure 3.1 Skull and dental evolution in the lineage leading to modern mammals.

Background

Teeth are extremely important to mammals (and to mammalogists). Modifications in the dentition and masticatory apparatus appear early in the diversification of mammals from their therapsid ancestors (Figure 3.1). These alterations reflect changes in physiology and behavior of ancestral mammals. Early mammals were small and probably foraged during the cooler hours from evening to dawn. Their small body size and nocturnal habits favored a consistently higher body temperature, which in turn required the ability to capture, chew, and digest food efficiently. The lineage leading to modern mammals is characterized by three important trends in dentition:

- reduction in the total number of teeth.
- reduction in the number of generations of teeth (e.g., from polyphyodonty to diphyodonty or monophyodonty).
- increased morphological complexity of teeth within the tooth row (e.g., heterodonty).

As the Mammalia continued to diversify, some of these trends were reversed; some odontocete whales and dolphins have more than 200 teeth, and those teeth are secondarily homodont. A few mammal species have lost teeth altogether. For most mammals, however, teeth have become variously specialized for snipping, grinding, crushing, puncturing, or slicing food. These diverse ways of processing food are reflected in the morphology of the teeth (Feldhamer et al., 2015; Vaughan et al., 2015). For example, the molars of herbivores are flattened and the occlusal surface is used to grind plant material. In contrast, the molars of carnivores (e.g., the hypercarnivorous felids) are blade-like and used to slice meat into more manageable pieces prior to swallowing. Indeed, we can often deduce the diet of the animal from its teeth alone. Finally, much of what we know about the evolutionary history of mammals we learned from teeth. This is because teeth are exceedingly hard and fossilize better than any other body part. Indeed, many species of fossil mammals are known only from their teeth. Fortunately, although teeth vary widely between species, they show very little variation within species. This feature makes them useful in reconstructing the phylogenetic history of mammals.

Internal Structure

Regardless of their shape, teeth share a common internal structure. The surface, or crown, is covered with a

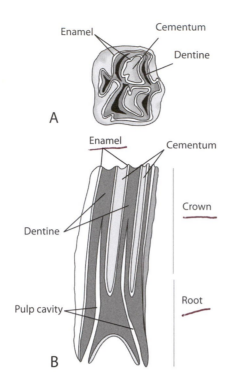

Figure 3.2
(**A**) Occlusal view of the surface of a horse molar showing the relationship of the enamel ridges and the depressions formed by the softer dentine. (**B**) Sagittal section through a horse molar, showing the internal structures of the tooth.

very hard substance called enamel. Just below the enamel is a layer of dentine. Dentine is hard, but not as hard as enamel. This differential hardness may result in surface ridges on the tooth, where the enamel and dentine are both exposed at different heights on the tooth's surface (Figure 3.2). The layer of dentine surrounds the pulp cavity. Pulp is a cellular tissue containing odontoblast cells, blood vessels, and nerves. The vessels and nerves enter the tooth via a canal at the base of the tooth (e.g., the root). The odontoblast cells in the pulp secrete dentine. The tooth is anchored in a socket in the jaw with a material called cementum. Despite the name, cementum does not act as a glue (sitting on the surface) but instead contains tiny connective tissue fibers rooted to the bone. In some species cementum may also extend onto the crown of the tooth.

In most mammals, the root contains a canal through which the blood vessels and nerves enter the pulp cavity from below. When the root canal is open, the tooth continues to grow, but when the animal matures, the root canal typically closes and the tooth stops growing. Rodent incisors are one example of teeth that continue to grow throughout the rodent's life (i.e., root-less or ever-growing teeth).

Kinds of Teeth

Teeth are present on three bones in mammals: the maxilla and premaxilla of the upper jaw, and the dentary of the lower jaw (Figure 3.3).

The ancestral homodont dentition has given way to four distinct kinds of teeth in many mammals: incisors, canines, premolars, and molars.

Incisors are the most anterior teeth, located in the premaxilla of the upper jaw and in the dentary of the lower jaw. Incisors are generally simple teeth, used for grasping and clipping. In many mammals, incisors have been modified. As mentioned, rodents have chisel-shaped, ever-growing incisors used for gnawing. Lemurs have modified incisors into a "toothcomb" used for grooming. Perhaps the most spectacularly modified incisors are the tusks of elephants.

Canines are posterior to the incisors. Typically, a single canine is present in each quadrant, and it is the first tooth in the maxilla. Canines are simple, conical teeth with a single cusp and a single root. They are used for stabbing or piercing. Rodents, many artiodactyls, and a few other taxa lack canines or have canines only in the lower jaw. Narwhals, musk deer, and the extinct saber-toothed cats have enlarged canines that probably serve other functions beyond feeding (e.g., social signaling, serving as defensive weapons).

Premolars lie immediately posterior to the canines in the maxilla and dentary. Premolars are deciduous, whereas molars are not replaced. Premolars vary from tiny pegs (shrews) to the massive crushing teeth of many herbivores (Figures 3.4 and 3.5).

Molars are the most posterior teeth in the jaws. They also vary tremendously in size and shape but exist only as adult teeth.

Mammalogists use a number of terms to refer to the orientation of tooth surfaces. The occlusal surface is the exposed surface that meets with the tooth in the opposite tooth row. The labial side of the tooth is adjacent to the lips, and the lingual side is adjacent to the tongue. Anterior and posterior have their traditional meaning when referring to teeth.

While mammalian ancestors evolved a complex series of teeth, each suited to a specialized function, some mammals have secondarily evolved homodont teeth. Homodont teeth are all similar in shape. In modern

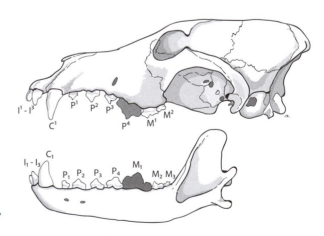

Figure 3.3 A dog skull illustrating a typical complement of teeth. The shaded upper premolar 4 and the lower molar 1 represent the carnassial pair in many carnivores.

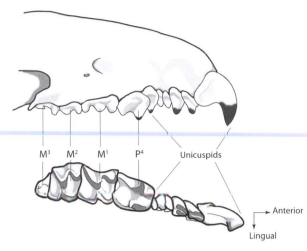

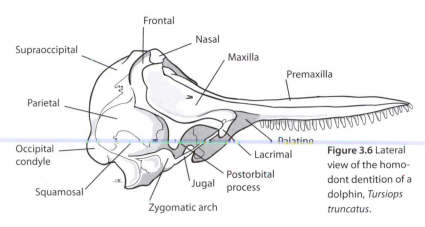

Figure 3.6 Lateral view of the homodont dentition of a dolphin, *Tursiops truncatus*.

Figure 3.4 Rostrum of a shrew, *Sorex cinereus*. The incisors, canines, and anterior premolars are called unicuspid teeth.

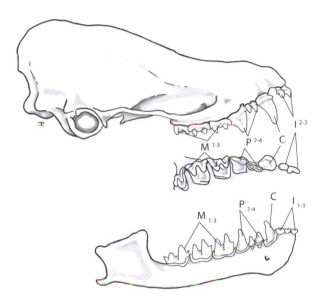

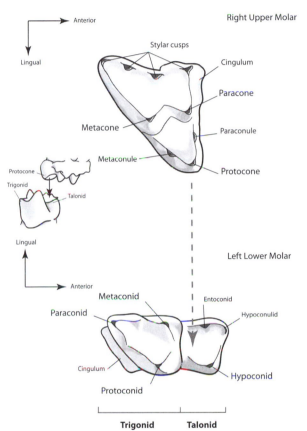

Figure 3.5 Skull and lower jaw of a little brown bat, *Myotis lucifugus*.

Figure 3.7 Occlusal views of a right-upper and left-lower tribosphenic molar. Notice that the protocone of the upper tooth fits into the basin of the talonid portion of the lower tooth.

mammals, homodont teeth tend to be found in fully aquatic piscivores (fish-eaters) such as toothed whales (Figure 3.6) and several species of pinnipeds.

Occlusal Patterns and Cusps

The occlusal surface of an erupted tooth has patterns of cusps and lophs that are species specific and often highly informative. Cusp morphology and position are correlated with diet and are the result of evolutionary and developmental constraints.

Early mammals had teeth with three major cusps in a relatively straight line (see Vaughan et al., 2015, for details). These cusps increased the slicing surfaces and provided an early form of mastication. Later, the central cusp was shifted to one side to form a triangle. This al-

lowed even greater occlusal surfaces for grinding and a more advanced form of mastication. The cheek teeth (molars and premolars) in particular often have a very complex topography. Mammalogists have created a vast array of names for the cusps, ridges, depressions, and other features of the occlusal surfaces of these teeth (Figure 3.7). While the full range of terminology is beyond the scope of this lab exercise, it is important for future mammalogists to understand the basic features of mammalian cheek teeth.

The first rule in dental terminology is that the cusps of the upper and lower teeth are given different suffixes. The upper cusps are designated with the suffix *-cone*; the three main cusps are the paracone, protocone, and

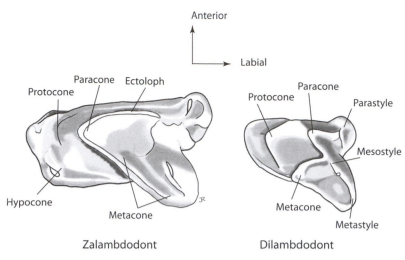

Figure 3.8 Occlusal views of zalambdodont and dilambdodont teeth.

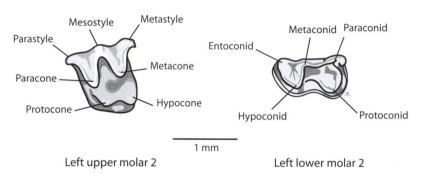

Figure 3.9 Occlusal views of the upper and lower molars of a little brown bat, *Myotis lucifugus*.

metacone. Secondary cusps in the upper teeth are designated by the suffix *-conule* (e.g., paraconule). In contrast, cusps of the lower teeth are designated with the suffix *-conid*, and secondary cusps use *-conulid*. Thus, the hypoconid is a cusp in the lower molar, and hypoconulid is a nearby secondary cusp (Figure 3.7). A cingulum is a shelf-like ridge around the outside of a molar (often called a cingulid for a lower molar). The stylar shelf is part of the cingulum that has become expanded and often bears several small cusps. A crista (or cristid) is a ridge, and a loph is a ridge formed by the fusion or expansion of cusps.

Early therians (placentals and marsupials) had cheek teeth similar to those illustrated in Figure 3.7. The upper molars were essentially triangular with three major cusps. The triangle is positioned with one side along the labial (lips) side of the tooth row and one apex pointing lingually (toward the tongue). The protocone is positioned at the lingual apex, the paracone is positioned at the anterior margin of the tooth, and the metacone is at the posterior margin. Early mammals also had a prominent stylar shelf (labial to the paracone and metacone). Because of its triangular shape, the upper cusps are said

to form a trigon. Lower molars in these early mammals contained two regions: a triangular trigonid and an attached lower region called the talonid. The cusps in the trigonid are reversed (relative to those in the upper molars). Here the protoconid is on the labial side and the paraconid and metaconid are on the lingual side of the jaw. The talonid contains a shallow basin surrounded by three cusps and is always on the posterior of the tooth (Figure 3.7). Because upper and lower "triangles" are reversed, the upper and lower teeth occlude (meet during chewing) precisely. The upper protocone fits snugly into the talonid basin of the lower tooth. Such tribosphenic teeth were common in early mammals but are found in only a few modern mammals, including the Virginia opossum (*Didelphis virginiana*).

Tribosphenic teeth have been modified in a variety of ways in the lineage leading to modern mammals as their diets diversified. For example, both zalambdodont and dilambdodont teeth retain a triangular shape (Figure 3.8). In zalambdodont upper molars there is a V-shaped crest (ectoloph) with the largest cusp, the paracone, at the apex. There is also an expanded stylar shelf, and the protocone is usually absent. Among living mammals, zalambdodont teeth are found in golden moles (family Chrysochloridae) and solenodons (family Solenodontidae).

Dilambdodont upper molars are similar but have a prominent W-shaped ectoloph (Figure 3.9), with the paracone and metacone at the bottom of the W.

The ectoloph crests extend labially to the stylar shelf. Lingual to the ectoloph is the protocone. Dilambdodont molars are characteristic of an insectivorous diet and are found in living soricid shrews (family Soricidae), moles (family Talpidae), and many insectivorous bats (Figures 3.4 and 3.5).

Later in mammalian evolution a fourth cusp, the hypocone, was added lingual and posterior to the protocone (Figure 3.10). This additional cusp squared the molar, forming a quadrate molar (also known as quadritubercular). Quadrate teeth are found in many mammals including hedgehogs (family Erinaceidae), raccoons (family Procyonidae), and many primates (e.g., family Hominidae). Additional modifications, such as crosslophs, occurred throughout the evolutionary history of mammals.

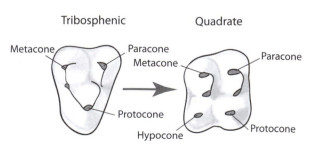

Figure 3.10 The transition from a tribosphenic to a quadrate upper molar by the addition of a hypocone.

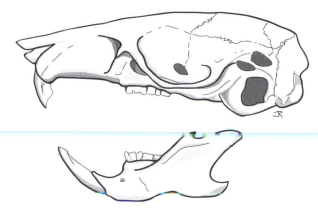

Figure 3.11 Lateral view of the skull of a deer mouse, *Peromyscus*, showing the ever-growing incisors and diastema.

Types of Teeth and Diet

Mammals feed on a wide variety of foods. Those that feed primarily on soft, less-abrasive materials (e.g., meat) have low-crowned teeth, termed brachydont. In contrast, herbivores that feed on grasses and other abrasive foods have hypsodont (high-crowned) teeth. These teeth are subject to greater wear and therefore extend well above the gum line. In some species, these hypsodont teeth are ever growing.

切齿

Carnivores slice meat using blade-like cheek teeth (Figure 3.3) referred to as secodont teeth. The fourth upper premolar and first lower molar in the jaws form a pair of carnassial teeth used to slice meat (not all members of the Carnivora have carnassials). Many carnivores also have large canines for stabbing and holding prey.

Piscivores eat a diet primarily of fish. Their teeth tend to be simple and conical in shape (secondarily homodont). This type of dentition is found in several species of seals and in odontocete cetaceans (Figure 3.6).

Insectivorous mammals eat hard and/or soft-bodied insects and other invertebrates. Their teeth are typically brachydont and either zalambdodont or dilambdodont (Figures 3.4 and 3.5). The prominent V- and W-shaped ectolophs form shearing or crushing zones for breaking up the hard chitinous exoskeletons of invertebrate prey. In a few species, such as armadillos, that specialize in soft-bodied prey, the teeth are reduced to simple pegs. Monotremes (Monotremata), Old World pangolins (Pholidota), and New World anteaters (Pilosa, Myrmecophagidae) have lost teeth. Instead they rely on a long, protrusible tongue covered in sticky mucus to lap up termites and ants.

Quadrate molars with rounded cusps are characteristic of many omnivorous mammals (Figure 3.10). Omnivores tend to have varied diets consisting of a mix of hard and soft foods including both plant and animal materials. In omnivores, the four primary cusps are blunt (bunodont) and the occlusal surfaces of the upper and lower molars oppose one another directly. Bunodont teeth are usually also brachydont (low crowned). Hominid

丘齿

primates, pigs (family Suidae), bears (family Ursidae), raccoons (family Procyonidae), and a few rodents (family Sciuridae) are all examples of omnivores with bunodont dentitions. Many of these species retain relatively well-developed canines.

Herbivores consume a diet of leaves, seeds, fruits, grasses, and other plant material. Different plants, and even different regions of the same plant, vary greatly in abrasiveness and digestibility. Therefore, it is not surprising that herbivores have evolved many types of teeth for dealing with the challenges of an herbivorous diet.

Rodents and rabbits have ever-growing incisors used for gnawing (Figure 3.11). The incisors are chisel-shaped and strongly beveled at the tips. The anterior surface of rodent incisors is often pigmented; the enamel here is harder than the dentine on the posterior surface, resulting in differential wear and the characteristic beveled shape of the tips. Rodents have a single incisor in each quadrant (one pair of upper and one pair of lower incisors). Rabbits have two pair of upper incisors, but the second pair is small and located behind the first pair. Gnawing wears the incisors away at the tips, but they are rootless and grow continuously throughout the animal's life.

Behind the incisors is the diastema, a gap where no teeth occur. The cheek teeth (premolars and molars) are typically hypsodont, flat, and have surface ridges and lophs for grinding plants. A wide variety of occlusal surfaces have evolved in rodents (Figures 3.12 and 3.13). For example, many squirrels (Sciurids) tend to have relatively simple brachydont cheek teeth, but voles and their relatives (Arvicolinae) have hypsodont teeth with complex crowns featuring folds of enamel and dentine (Figure 3.12).

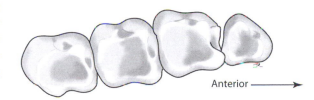

Sciurus aestuans

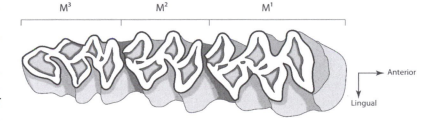

Ondatra zibethicus

Figure 3.12 Occlusal view of the cheek teeth of a Guianan squirrel, *Sciurus aestuans* (*top*) and a muskrat, *Ondatra zibethicus* (*bottom*).

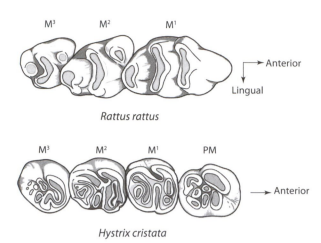

Figure 3.13 Occlusal view of the upper cheek teeth of a roof rat, *Rattus rattus* (*top*) and an African crested porcupine, *Hystrix cristata* (*bottom*).

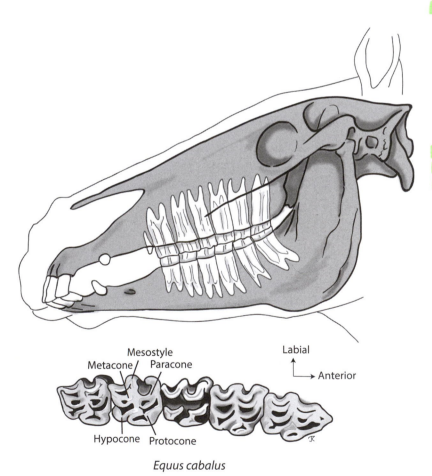

Equus cabalus

Figure 3.14 Lateral view of a horse head illustrating the elongate ever-growing cheek teeth set in deep sockets (*top*). Occlusal view of the right-upper tooth row of a horse (*bottom*).

In horses and other grazing mammals, the premolars tend to become molariform (appear molar-like). The cheek teeth have flattened surfaces with numerous exposed ridges of enamel and dentine. These ridges act to grind coarse grasses into finer particles that can be more easily digested. However, grazing animals also ingest a lot of dust and dirt along with the grasses they consume. Such an abrasive diet leads to rapid tooth wear; for this reason, grazing mammals typically have hypsodont teeth (Figure 3.14), or high-crowned teeth. Horse incisors, premolars, and molars continue to grow as the grinding surface is worn down. In an adult horse the majority of the tooth lies beneath the gum line in deep, bony sockets (Figure 3.14).

Deer, moose, bison, antelopes, and other members of the families Cervidae and Bovidae are browsing mammals. These mammals eat leaves, soft shoots, fruits, or buds of woody plants. Browsers often lack upper incisors and canines (Figure 3.15). Instead, these teeth are replaced with a hard pad on the palate, which in concert with the lower incisors (which are generally retained) clip foliage for further processing by the cheek teeth. There is a broad gap or diastema anterior to the cheek teeth. Browsers have low-crowned (brachydont) but selenodont cheek teeth. Selenodont teeth are functionally similar to lophodont teeth (discussed below), except that the enamel ridges are oriented in an anterior-posterior plane and form characteristic crescent-shaped cusps (Figure 3.15).

Browsing mammals may also have lophodont teeth. Lophodont teeth also have elongate ridges spanning the tooth, but in this case the lophs between cusps are perpendicular to the tooth row (labio-lingual). Many rodents (Rodentia) as well as tapirs (family Tapiridae) and manatees (family Trichechidae) have lophodont molars. Elephants (family Elephantidae) and capybaras (family Rodentia) exhibit an elaborate form of lophodonty called loxodont dentition (Figure 3.16). The lophs of loxodont teeth are numerous and form a series of closely-spaced ridges suitable for grinding hard, woody foods.

Although they technically lack teeth, baleen whales (Mysticete) have evolved a unique method of feeding worth mentioning here. In place of true teeth, baleen whales have racks of parallel rows of baleen in the upper jaw that serve as a sieve to filter out small crustaceans (e.g., krill and copepods) and small fish from the vast quantities of seawater they engulf during each feeding bout. Baleen is made of keratin (as are claws and hairs) and forms flat, flexible plates between 0.5 and 3.5 meters long (Figure 3.17). The distal end of each baleen plate is frayed. Two parallel rows of closely-spaced baleen plates run along the length of the upper jaw (for details see Vaughan et al., 2015). The comb-like rows of baleen form a sieve or filter upon which to capture tiny prey.

Different species of baleen whales feed in different manners (Vaughan et al., 2015), but they share a common

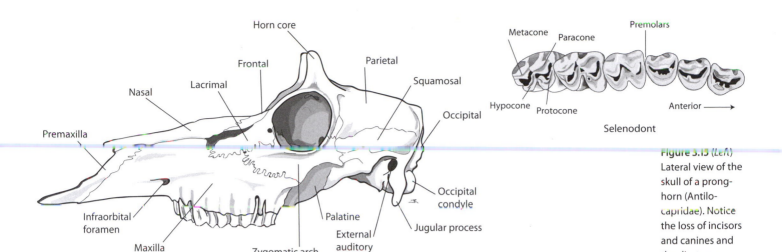

Selenodont

Figure 3.15 (*Left*) Lateral view of the skull of a prong-horn (Antilocapridae). Notice the loss of incisors and canines and the diastema anterior to the cheek teeth. (*Right*) Occlusal surface of the cheek teeth of a white-tailed deer. Notice the crescent-shaped ridges that run parallel with the tooth row.

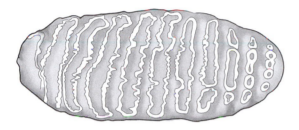

Elephas maximus

Figure 3.16 Occlusal view of the molar of an Asian elephant, *Elephus maximus*, showing the many crosslophs of a loxodont tooth.

Figure 3.18 Drawing of a humpback whale as it swims upward through a shoal of prey.

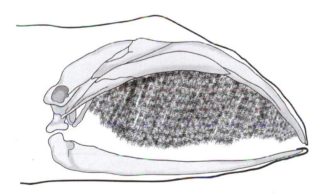

Figure 3.17 Lateral view of the skull of a baleen whale with the baleen plates intact.

means of filtering prey from the water. For example, humpback whales (*Megaptera novaeangliae*) swim upward through dense swarms of prey. As they near the surface, they open their mouths. The furrowed skin on the throat expands to form a vast "net" or scoop. As they swim through the school of prey, they engulf large quantities of prey just as they breach the surface (called lunge feeding). The whale closes its mouth and uses its tongue and the sieve-like baleen to force out the water, trapping the tiny prey against the rows of baleen. After the water is removed, the whale swallows its meal (Figure 3.18).

Tooth Replacement 多换性齿

Reptiles, and presumably the early ancestors of mammals, have polyphyodont teeth, or continual tooth replacement. As a tooth is lost, it is eventually replaced by another tooth, and the process continues throughout the animal's life. This leads to an uneven tooth row. In contrast, mammals have diphyodont dentition—they have 一换性 two sets of teeth: a deciduous set (milk teeth) is replaced by a permanent set of adult teeth. The milk teeth consist of incisors, canines, and premolars, but molars are part of the adult dentition and are not replaced. The main reason for mammals' reduced number of lifetime tooth replacements is that mammals need a full, even tooth row with teeth that are firmly attached in order to withstand the forces required for chewing and grinding food. Among

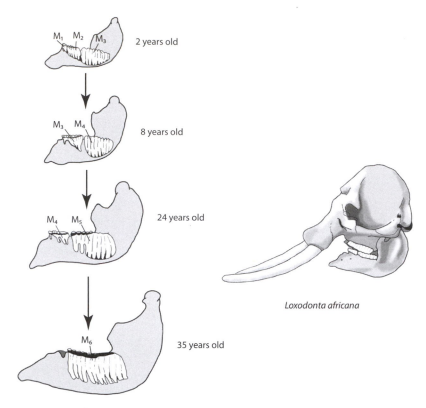

2 years old

8 years old

24 years old

35 years old

Loxodonta africana

Figure 3.19 Lateral views of the skull (*right*) lower jaw and molar teeth (*left*) in the African elephant (*Loxodonta africana*). The tusks in the skull on the right are modified incisors. The progression of jaws on the left shows the movement of molars forward in the jaw as the animal ages. *Adapted from Kingdon (1979)*

mammals, the timing and style of replacement vary. For example, odontocete cetaceans have homodont dentition and only one set of teeth is ever formed; these teeth must last a lifetime. In many rodents (and some pinnipeds), replacement of the deciduous teeth takes place before birth and the young are born with their adult teeth.

Kangaroos also have an unusual pattern of tooth replacement (Dawson, 1995). Kangaroos eat grass and other abrasive plant material, and their teeth are subject to considerable wear. Placental herbivores have teeth that grow continuously as they wear, but kangaroos do not have open-rooted, continuously growing cheek teeth. Instead they solve the problem by moving a new set of molars into grinding position along the tooth row as the previous set is worn away. Thus, the molars erupt in succession from back to front as the animal ages.

Manatees are also an exception to diphyodont dentition in mammals (Domning and Hayek, 1984). They only have molars, and these molars are replaced in progression from back to front. The so-called marching molars enter at the back of the tooth row and constantly move forward in the jaw. By the time the molar gets to the front it is well-worn and the roots degenerate, leading to the loss of the remnant of the tooth. Thus, a manatee has an indeterminate number of molars with up to seven molars per quadrant in various states of wear in its jaw at any point in time.

Elephants (related to manatees) also have molars that are replaced throughout life, but they have only a limited set of these replacement molars (this process is described in detail in Kingdon, 1979). African elephants (*Loxodonta africana*) have up to six molars per quadrant in their lower

jaw in a lifetime, but only one or two molars are ever fully erupted at the same time (Figure 3.19). Each molar erupts at the back of the jaw and moves forward with time. As the surface is exposed and worn away against the abrasive browse that elephants eat, the tooth is pushed farther forward by those from behind. As the tooth reaches the front, it is heavily worn and the roots begin to be reabsorbed. Eventually the tooth falls out or is swallowed. If the elephant reaches 40 years of age, it typically has its last (sixth) molar in place in the lower tooth row.

Dental Formulas

Dental formulas are simple ways to denote the complement of teeth in a species. Mammalogists typically use four quadrants when referring to the position of teeth; right upper jaw, left upper jaw, right lower jaw, and left lower jaw. Dental formulas describe the number and type of teeth in the upper and lower tooth rows of one side of the head (i.e., right upper and right lower tooth rows). Each tooth type (incisor, canine, premolar, molar) is specified along with the number of each type. There are several ways of expressing a dental formula, and to read the vast literature on dental evolution in mammals, it is important to understand each. However, most mammalogists use the dental formula format described here.

Consider the dental formula for a wolf (*Canis lupis*):

$$I\frac{3}{3} \; C\frac{1}{1} \; P\frac{4}{4} \; M\frac{2}{3}$$

This formula indicates that the wolf has three upper and three lower incisors followed by one upper and one lower canine, four upper and four lower premolars, and two upper and three lower molars. The letters denote the type of tooth, and the numbers represent the number of each kind of tooth in the upper ("numerator") and lower ("denominator") jaws. Totaling the "numerators" and "denominators" gives 10 upper teeth over 11 lower teeth, or 21 teeth on one side of the head. This number is doubled to get the total number of teeth in the adult animal (e.g., 42 teeth in the wolf).

Primitive metatherians and eutherians generally had more teeth than many living species. For example, the dental formula for a representative primitive metatherian is

$$I\frac{5}{4} \; C\frac{1}{1} \; P\frac{3}{3} \; M\frac{4}{4} = 50$$

During the evolutionary history of mammals, these numbers have often been reduced. The eastern cottontail rabbit (*Sylvilagus floridanus*) has a reduced number of incisors and lacks canines altogether; its dental formula is

$$I\frac{2}{1} \; C\frac{0}{0} \; P\frac{3}{2} \; M\frac{3}{3} = 28$$

Exercise 1: Dental Terminology

Your instructor will provide you with a set of mammal skulls with a letter code to identify each skull. For each skull, complete Table 3.1 by entering the dental formula, cheek tooth type, mandibular and maxillary tooth row length (in millimeters), and a hypothesized diet.

Recall the following definitions for the measurements from Chapter 2:

- **Length of maxillary tooth row**: distance from the anterior margin of the upper canine tooth to the posterior-most margin of the last molar on the maxilla.
- **Length of mandibular tooth row**: distance from the anterior margin of the lower canine tooth to the posterior-most margin of the last molar on the mandible.

Table 3.1 Data table for mammal skulls

Skull	Dental formula	Cheek tooth type (e.g., selenodont)	Length of mandibular tooth row (mm)	Length of maxillary tooth row (mm)	Hypothesized diet

Exercise 2: Dental Key to North American Mammals

Using the information provided in Table 3.1, develop a dichotomous key to the skulls your instructor provided. This key should use only the terminology and informa-tion described in this chapter (i.e., dental terms, tooth features, and measurements). Use Table 3.2 to construct your key.

Table 3.2 Dichotomous key worksheet

Couplet	Description	Go to

4   Phylogeny Reconstruction

Time Required
- One 3- to 4-hour lab period

Learning Objectives
- Understand how phylogenies are constructed.
- Understand how morphological characters are used to construct phylogenetic trees.
- Understand how molecular characters are used to build phylogenetic trees.
- Practice manual sequence alignment and tree construction.
- Use internet-based software tools to create and explore phylogenetic hypotheses.

Equipment Required
- Computers with access to the internet.

Background

Phylogenetics is the study of evolutionary relatedness among groups of organisms. The data used to discover these relationships may be molecular sequencing data and/or morphological data. Phylogenetics is related to, but distinct from, taxonomy—the naming and classification of organisms. Phylogenetic analyses are essential tools for elucidating the evolutionary tree of life (Figure 4.1). Evolution produced only one true phylogenetic tree of all living organisms. It is our job as scientists to recover that one true tree using the best available methods. Evolution is a change in one or more traits in a population over time. It is, therefore, regarded as a branching process, in which populations change over time. If they change sufficiently they may

- form separate lineages (branches) and eventually separate species,
- come together via hybridization, or
- terminate by extinction.

All of these processes may be mapped on a phylogenetic tree. A phylogenetic tree is a testable hypothesis of the evolutionary events. We can never go back in time and check a tree's accuracy, but we can test the tree's branching patterns by adding more or different data, thereby testing the phylogeny.

How do we produce a phylogenetic tree (hypothesis) in the first place? The current method to infer phylogenetic trees is called cladistics, which includes a number of more specific methods, such as parsimony, maximum likelihood, and Bayesian inference. All of these require complex mathematical algorithms to establish relationships, which are then represented graphically as phylogenetic trees.

Morphological characters can be used in the reconstruction of phylogenies. Today, though, molecular data, which include protein and DNA sequences, are the primary source of data for constructing phylogenetic trees. For a more complete discussion of phylogenetics, see Hall and Hallgrimsson (2008) or Hall (2006).

How Do We Construct Phylogenetic Trees?

Phylogenetic trees represent patterns of ancestry and are therefore similar to family trees (genealogy). However, while families may have kept records of births and other relationships, evolution does not. It falls to biologists to reconstruct those evolutionary histories by analyzing other types of evidence and using those data to form a phylogenetic hypothesis about how some group of organisms are related. The evidence biologists collect includes morphological or genetic characters of each organism in the group they are studying. Regardless of the type of character, all characters must be heritable traits that were passed down through the lineage. The goal is to find shared derived characters that will group organisms into monophyletic clades.

The steps in phylogenetic reconstruction are as follows:

- Choose a set of taxa of interest.
- Examine each taxon carefully and determine the characters that will be included in the analysis.
- Determine whether each character state is ancestral or derived for each taxon. This may require looking

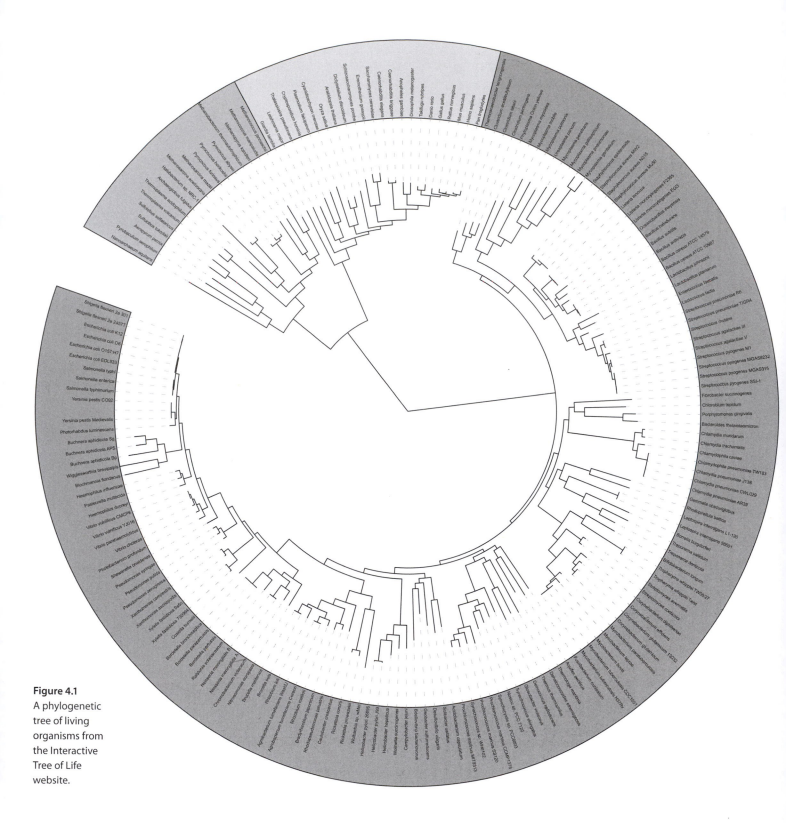

Figure 4.1
A phylogenetic tree of living organisms from the Interactive Tree of Life website.

at taxa outside of, but related to, the group being studied (e.g., an outgroup).

- Group taxa by shared derived characteristics (synapomorphies). Shared ancestral characters are not informative.
- Reduce potential conflicts in the data set using parsimony (minimizing the number of conflicts).
- Build the cladogram.

These basic steps for creating a valid phylogeny are simple and reasonably straightforward. However, the devil is in the details (as usual); constructing robust, well-supported phylogenies requires large data sets, complicated algorithms, and considerable time. To illustrate the process, we will work through a simple example using morphological characters. Suppose you are interested in the taxa in Table 4.1 and you have examined

Table 4.1 Morphological characters for six vertebrates

Morphological character	Iguana	Platypus	Kangaroo	Baboon	Chimp	Vampire bat
Four legs	X	X	X	X	X	X
Hair		X	X	X	X	X
Mammary glands		X	X	X	X	X
Marsupium			X			
Chorioallantoic placenta				X	X	X
Wings						X
Echolocation						X
Stereoscopic vision				X	X	
Opposable thumbs				X	X	
Viviparity			X	X	X	X

them for 10 characters. In Table 4.1 the presence of a character is denoted by an X.

How could we use this data matrix to construct a reasonable phylogeny? The first step is to determine which character(s) are the most widely shared. In this matrix, all six taxa share the presence of four legs (i.e., they are all tetrapods). This character is uninformative in our quest to establish the evolutionary history of the six taxa listed because they all share it; this character must have evolved prior to the evolution of the six taxa. The next most widely shared characters are hair and mammary glands (shared by five of six taxa). Iguanas lack both hair and mammary glands. Therefore, iguanas must have branched off prior to the lineage containing the remaining taxa. We could illustrate this by a tree diagram like the one shown in Figure 4.2.

We still do not know how the platypus, kangaroo, chimp, baboon, and vampire bat are related to one another. Analysis of the next most widely shared character may help. Four of the remaining five taxa share viviparity (i.e., they give birth to live young). The platypus lays eggs so it must have branched off from the remaining taxa prior to the evolution of viviparity. The new phylogeny is now somewhat better resolved as shown in Figure 4.3.

As we continue analyzing each set of shared characters, we resolve the relationships of the taxa. The next most widely shared character is having a chorioallantoic placenta. Kangaroos have a simpler choriovitelline placenta. Thus, we can unite the chimps, baboon, and vampire bats in a group based on the shared presence of the chorioallantoic placenta. Kangaroos must have branched from the remaining three taxa prior to the evolution of this specialized placenta (Figure 4.4).

The next most widely shared characters are opposable thumbs and stereoscopic vision, both of which are found in chimps and baboons but not in vampire bats. There are also several characters that are unique to one

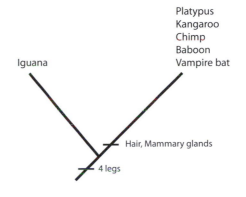

Figure 4.2 Step one in the reconstruction of the phylogeny from data in Table 4.1.

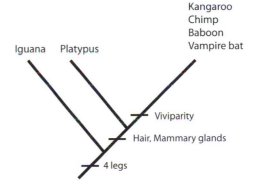

Figure 4.3 The platypus, a monotreme, branched off prior to the evolution of viviparity.

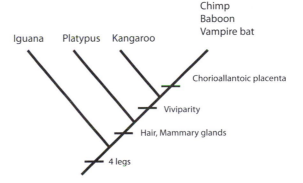

Figure 4.4 Kangaroos, members of the Metatheria, branch off next.

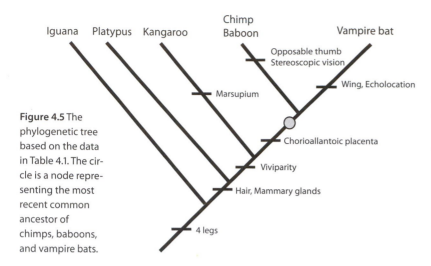

Figure 4.5 The phylogenetic tree based on the data in Table 4.1. The circle is a node representing the most recent common ancestor of chimps, baboons, and vampire bats.

taxon. These are placed on the line leading to that taxon as shown in the fully resolved tree in Figure 4.5.

The taxa go on the end of the branches. Nodes lie where two or more branches come together (white dot in Figure 4.5); a node represents the hypothetical most recent common ancestor for all taxa above that node. For example, the node in Figure 4.5 would be the most recent common ancestor of chimps, baboon, and vampire bats (eutherian mammals). A list of synapomorphies that are common to all taxa above the node are usually indicated by a horizontal bar. Of course, we can also use other kinds of data, such as molecular sequence data, to construct cladograms.

Exercise 1: Manual Sequence Alignment

In this exercise, you will practice two methods of determining the evolutionary relationships between organisms: molecular similarity and cladistic analysis. Classifying based on molecular similarity uses comparisons of the amino acid sequence differences in proteins to indicate relatedness. This method has the advantage that it is relatively easy to understand. Highly related organisms are expected to have many sequence similarities; distantly related animals are expected to have low similarity. You will draw conclusions about the relationships of the same six taxa shown in Figure 4.5: common iguana (*Iguana iguana*), duck-billed platypus (*Ornithorhynchus anatinus*), eastern gray kangaroo (*Macropus giganteus*), olive baboon (*Papio anubis*), chimpanzee (*Pan troglodytes*), and the false vampire bat (*Megaderma lyra*). How do you predict the resulting phylogeny will look? What do you predict about the molecular similarity of organisms within each pair?

Each amino acid in the protein chain is represented by a unique single-letter code assigned by the International Union of Pure and Applied Chemistry (IUPAC) (Table 4.2).

The first step is a laborious one: you must find the number of differences in amino acid sequence between each pair of species, using the data in the file "Exercise 4.1 Alignment data" available from the website that accompanies this manual (jhupbooks.press.jhu.edu/content /mammalogy-techniques-lab-manual). A portion of the data is shown in Table 4.3. There you can see that, for

Table 4.2 Single-letter IUPAC codes for the 20 standard amino acids in alphabetical order

A alanine	G glycine	M methionine	S serine
C cysteine	H histidine	N asparagine	T threonine
D aspartic acid	I isoleucine	P proline	V valine
E glutamic acid	K lysine	Q glutamine	W tryptophan
F phenylalanine	L leucine	R arginine	Y tyrosine

Table 4.3 A sample of the amino acid sequence data from the downloaded data set

Sequence number	Iguana	Platypus	Kangaroo	Baboon	Chimp	Vampire bat
1	V	V	V	V	V	V
2	H	H	H	H	H	H
3	W	L	L	L	L	L
4	V	T	S	T	T	T
5	A	G	A	P	P	N
6	E	G	E	E	E	E
7	E	E	E	E	E	E
8	K	K	K	K	K	K
9	Q	S	N	S	N	T
10	L	A	A	A	A	A

Table 4.4 Pair-wise combinations for amino acid differences between species for the hemoglobin beta chain

	Iguana	Platypus	Kangaroo	Baboon	Chimp	Vampire bat
Iguana						
Platypus						
Kangaroo						
Baboon						
Chimp						20
Vampire bat						

example, all six taxa share the amino acid valine (V) at position 1, but the iguana has a tryptophan (W) at position 3 whereas the other five taxa have a leucine (L) at that position. There are 15 species pairs to be analyzed, as shown by the unshaded cells in Table 4.4. To complete the table, compare two taxa at a time and record the number of amino acid differences between them. For example, there are 20 places in the chain where there is an amino acid difference between the chimp and the vampire bat.

Divide the pair-wise comparisons up among group members so that each person is responsible for a certain subset of pair-wise comparisons. Enter the number of amino acid differences for each pair of species in the Table 4.4 (the chimp-vampire pair-wise comparisons is already entered for you). Using the data from Table 4.4, construct a phylogenetic tree for these six taxa. Before you proceed, answer the following questions:

- According to the data, which two species in Table 4.4 are most closely related?
- Why do you think these two lineages diverged the shortest time ago?
- Which species are the most distantly related?
- How does the molecular phylogeny compare with the phylogeny based on morphological characters (Figure 4.5)?

Exercise 2: Sequence Alignment Using Computers

In this exercise, you will construct a molecular phylogeny based on the same hemoglobin beta chain protein. Instead of doing this by hand, you will use the power of computers to do the sequence alignment. Use the UniProt database website at http://www.uniprot.org (UniProt Consortium, 2014; Figure 4.6).

Figure 4.6 The UniProt protein database home page.

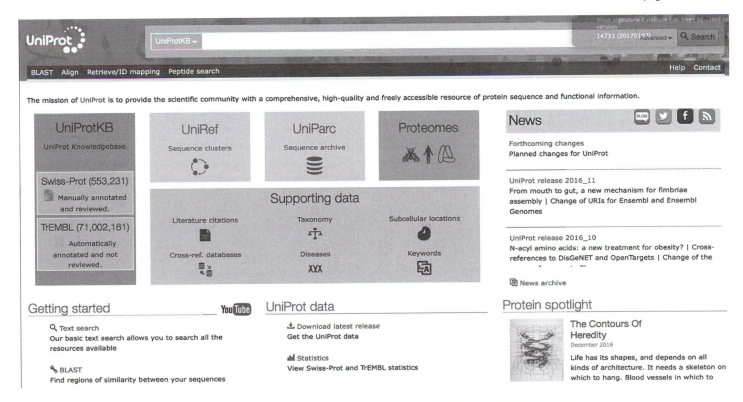

UniProtKB results

Filter by[i]

📄 Reviewed (194)
Swiss-Prot

📄 Unreviewed (330)
TrEMBL

Popular organisms

Human (64)

Mouse (28)

Rat (9)

Bovine (6)

RABIT (4)

Other organisms

[_____] Go

Search terms

Filter "hbb" as:

gene name (432)

protein name (36)

View by

[Results table]

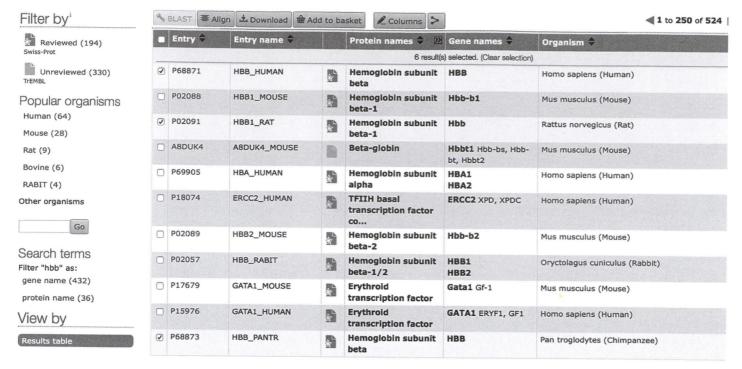

🔍 BLAST ≋ Align ⬇ Download 🗺 Add to basket ✎ Columns ＞   ◀ 1 to 250 of 524 |

☐	Entry ⬍	Entry name ⬍		Protein names ⬍ ⏩	Gene names ⬍	Organism ⬍
				6 result(s) selected. (Clear selection)		
☑	P68871	HBB_HUMAN	📄	Hemoglobin subunit beta	HBB	Homo sapiens (Human)
☐	P02088	HBB1_MOUSE	📄	Hemoglobin subunit beta-1	Hbb-b1	Mus musculus (Mouse)
☑	P02091	HBB1_RAT	📄	Hemoglobin subunit beta-1	Hbb	Rattus norvegicus (Rat)
☐	A8DUK4	A8DUK4_MOUSE	📄	Beta-globin	Hbbt1 Hbb-bs, Hbb-bt, Hbbt2	Mus musculus (Mouse)
☐	P69905	HBA_HUMAN	📄	Hemoglobin subunit alpha	HBA1 HBA2	Homo sapiens (Human)
☐	P18074	ERCC2_HUMAN	📄	TFIIH basal transcription factor co...	ERCC2 XPD, XPDC	Homo sapiens (Human)
☐	P02089	HBB2_MOUSE	📄	Hemoglobin subunit beta-2	Hbb-b2	Mus musculus (Mouse)
☐	P02057	HBB_RABIT	📄	Hemoglobin subunit beta-1/2	HBB1 HBB2	Oryctolagus cuniculus (Rabbit)
☐	P17679	GATA1_MOUSE	📄	Erythroid transcription factor	Gata1 Gf-1	Mus musculus (Mouse)
☐	P15976	GATA1_HUMAN	📄	Erythroid transcription factor	GATA1 ERYF1, GF1	Homo sapiens (Human)
☑	P68873	HBB_PANTR	📄	Hemoglobin subunit beta	HBB	Pan troglodytes (Chimpanzee)

Figure 4.7 The list of all HBB sequences in the UniProtKB database for mammals.

Across the top of the web page is a search bar. Enter the abbreviation for hemoglobin beta chain as **hbb** in the search bar, and click the Search button on the right. This takes you to a list of species that have known amino acid sequences for hemoglobin beta chain. You can select the species you want to analyze by checking the box to the left of the entry (Figure 4.7)

For this exercise, you will construct a phylogeny of primates. In the search bar, enter **hbb primates** and click Search. The new search page shows hemoglobin sequences only for primate species. Choose 10 primate species to use in your analysis by clicking in the box to the left of each entry. When you have chosen the 10 species you want to consider, click the Align button. The program aligns all the sequences and constructs a phylogenetic tree based on the number of sequence differences, just as you did manually (Figures 4.8 and 4.9). Click the Download button above the sequence data. Use the uncompressed FASTA format in the popup window. Copy and paste this phylogenetic tree to a text editor on your computer to save it, and write a two-page summary of the analysis.

Choose another set of taxa (not primates) and repeat the hemoglobin sequence analysis, or repeat the same primate analysis using a different protein (i.e., cytochrome c). You can select taxa and save them to your basket so that you do not have to hunt for each one again later. To do this, click the box to the left of each entry and then click the Add to Basket tab at the top of the page.

Alignment

🖶 How to print an alignment in color

```
P68871 HBB_HUMAN    1  MVHLTPEEKSAVTALWGKVNVDEVGGEALGRLLVVYPWTQRFFESFGDLSTPDAVMGNPK   60
P68873 HBB_PANTR    1  MVHLTPEEKSAVTALWGKVNVDEVGGEALGRLLVVYPWTQRFFESFGDLSTPDAVMGNPK   60
Q9TSP1 HBB_PAPAN    1  MVHLTPEEKNAVTALWGKVNVDEVGGEALGRLLVVYPWTQRFFDSFGDLSSPAAVMGNPK   60
P02050 HBB_OTOCR    1  -VHLTPDEKNAVCALWGKVNVEEVGGEALGRLLVVYPWTQRFFDSFGDLSSPSAVMGNPK   59
P02110 HBB_TACAC    1  MVHLSGSEKTAVTNLWGHVNVNELGGEALGRLLVVYPWTQRFFESFGDLSSADAVMGNAK   60
P02036 HBB_SAISC    1  MVHLTGDEKAAVTALWGKVNVEDVGGEALGRLLVVYPWTQRFFESFGDLSTPDAVMNNPK   60
P02024 HBB_GORGO    1  MVHLTPEEKSAVTALWGKVNVDBVGGEALGRLLVVYPWTQRFFESFGDLSTPDAVMGNPK   60
P68224 HBB_MACSP    1  MVHLTPEEKNAVTTLWGKVNVDEVGGEALGRLLVVYPWTQRFFESFGDLSSPDAVMGNPK   60
P02033 HBB_PILBA    1  -VHLTPDEKNAVTALWGKVNVDEVGGEALGRLLVVYPWTQRFFDSFGDLSTADAVMGNPK   59
Q6WN21 HBB_CALGO    1  MVHLTGEEKSAVTTLWGKVNVDEVGGEALGRLLVVYPWTQRFFESFGDLSSPDAVMNNPK   60
P08259 HBB_MANSP    1  -VHLTPEEKTAVTTLWGKVNVDEVGGEALGRLLVVYPWTQRFFDSFGDLSSPDAVMGNPK   59
                       ***: .** ** ***:***::*****************:******:   ***.* *


P68871 HBB_HUMAN   61  VKAHGKKVLGAFSDGLAHLDNLKGTFATLSELHCDKLHVDPENFRLLGNVLVCVLAHHFG  120
P68873 HBB_PANTR   61  VKAHGKKVLGAFSDGLAHLDNLKGTFATLSELHCDKLHVDPENFRLLGNVLVCVLAHHFG  120
Q9TSP1 HBB_PAPAN   61  VKAHGKKVLGAFSDGLNHLDNLKGTFAQLSELHCDKLHVDPENFKLLGNVLVCVLAHHFG  120
P02050 HBB_OTOCR   60  VKAHGKKVLSAFSDGLQHLDNLCGTFAKLSELHCDKLHVNPENFRLLGNVLVCVLAHHFG  119
P02110 HBB_TACAC   61  VKAHGAKVLTSFGDALKNLDNLKGTFAKLSELHCDKLHVDPENFNRLGNVLVVVLARHFS  120
P02036 HBB_SAISC   61  VKAHGKKVLGAFSDGLAHLDNLKGTFAQLSELHCDKLHVDPENFRLLGNVLVCVLAHHFG  120
P02024 HBB_GORGO   61  VKAHGKKVLGAFSDGLAHLDNLKGTFATLSELHCDKLHVDPENFRLLGNVLVCVLAHHFG  120
P68224 HBB_MACSP   61  VKAHGKKVLGAFSDGLNHLDNLKGTFAQLSELHCDKLHVDPENFKLLGNVLVCVLAHHFG  120
P02033 HBB_PILBA   60  VKAHGKKVLGAFSDGLAHLDNLKGTFAQLSELHCDKLHVDPENFKLLGNVLVCVLAHHFG  119
Q6WN21 HBB_CALGO   61  VKAHGKKVLGAFSDGLTHLDNLKGTFAQLSELHCDKLHVDPENFRLLGNVLVCVLAHHFG  120
P08259 HBB_MANSP   60  VKAHGKKVLGAFSDGLNHLDNLKGTFAQLSELHCDKLHVDPENFKLLGNVLVCVLAHHFG  119
                       ***** *** :*.*.* :**** **** ***********:****. ****** ***:**.


P68871 HBB_HUMAN  121  KEFTPPVQAAYQKVVAGVANALAHKYH                                   147
P68873 HBB_PANTR  121  KEFTPPVQAAYQKVVAGVANALAHKYH                                   147
Q9TSP1 HBB_PAPAN  121  KEFTPQVQAAYQKVVAGVANALAHKYH                                   147
P02050 HBB_OTOCR  120  KDFTPEVQAAYEKVVAGVATALAHKYH                                   146
P02110 HBB_TACAC  121  KEFTPEAQAAWQKLVSGVSHALAHKYH                                   147
P02036 HBB_SAISC  121  KEFTPQVQAAYQKVVAGVANALAHKYH                                   147
P02024 HBB_GORGO  121  KEFTPPVQAAYQKVVAGVANALAHKYH                                   147
P68224 HBB_MACSP  121  KEFTPQVQAAYQKVVAGVANALAHKYH                                   147
P02033 HBB_PILBA  120  KEFTPQVQAAYQKVVAGVANALAHKYH                                   146
Q6WN21 HBB_CALGO  121  KEFTPTVQAAYQKVVAGVANALAHKYH                                   147
P08259 HBB_MANSP  120  KEFTPQVQAAYQKVVAGVANALAHKYH                                   146
                       *:*** .***::*:*:**: *******
```

Figure 4.8 The HBB (hemoglobin) sequences alignment page showing the 146–147 amino acid sequences aligned.

Tree

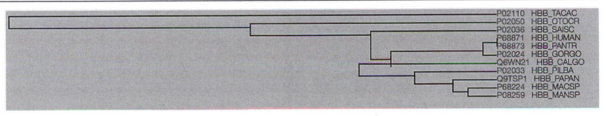

☐Highlight Taxonomy

Figure 4.9 The phylogenetic tree resulting from the HBB sequence data shown in Figure 4.8.

Exercise 3: Exploring the Open Tree of Life

The Open Tree of Life is an online tool for displaying and exploring phylogenetic trees (Hinchliff et al., 2015). It provides access to a wide array of published trees. The tree data can be viewed or downloaded for entry into other phylogenetic tree programs, such as the interactive Tree of Life (Letunic and Bork, 2016).

Let's explore the Open Tree of Life looking for mammalian phylogenies.

1. Go to https://tree.opentreeoflife.org.

2. In the Search for Taxon box at the top of the page, enter the word **Mammalia** (Figure 4.10), and hit return.

3. The search returns a tree like that shown in Figure 4.11. Notice that the nodes are indicated by gray circles. Clicking on these nodes will further refine that level of the phylogeny.

4. Click on the Eutheria node and explore the phylogeny of that group. As you continue to explore the tree,

eventually you get to the species-level trees for each genus (Figure 4.12).

5. You can also go to the data on the published study that supports the given phylogenetic tree by clicking on the name of the author in the list of supporting trees on the lower right of the window. There are tabs across the top for Metadata and Trees, among other things (Figure 4.13).

6. Click on the Tree tab and then on one of the trees listed on the left side of the table. It pulls up the tree as discussed in that study.

7. Download that tree configuration by clicking on the Metadata tab and on the Newick button in the Download section (Figure 4.13 at lower right). This downloads the tree in Newick format.

8. Copy the Newick format data to a text editor or clipboard and save it.

Figure 4.10 The Open Tree of Life website home page.

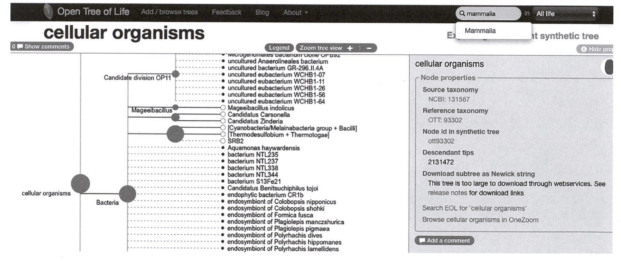

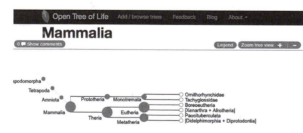

Figure 4.11 The Open Tree of Life website showing the results of a search for Mammalia.

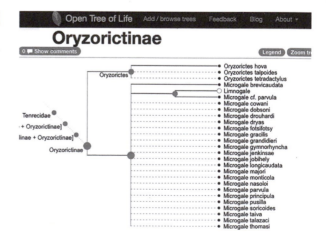

Figure 4.12 Open Tree of Life website showing a species-level tree for tenrecs.

9. Next, open the Interactive Tree of Life at http://itol.embl.de.

10. Click on the Upload tab at the top of the screen, and paste your Newick tree file into the Tree Text box (Figure 4.14). Give the tree an appropriate name, and click the Upload button at the bottom. The Newick tree format represents the tree as a series of nested parentheses and commas. It makes the tree readable by computers.

11. Your tree will appear on the main iTOL screen. Here you can change the look of the tree using the tools on the left. For example, you can replot it using the Circular, Normal, or Unrooted views. You can also explore the taxa on the tree by moving the cursor over each node in the tree.

Practice using these informative sites by uploading different subgroups of mammals and printing their phylogeny. Now that you have a basic understanding of how the Open Tree of Life and Interactive Tree of Life sites work, construct a tree of a mammal group using both sites and write a four- to five-page paper explaining the evolution of the group.

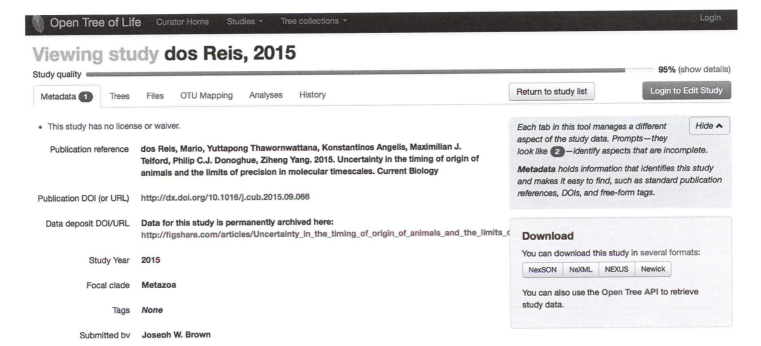

Figure 4.13 The Open Tree of Life web page for author information and downloading data.

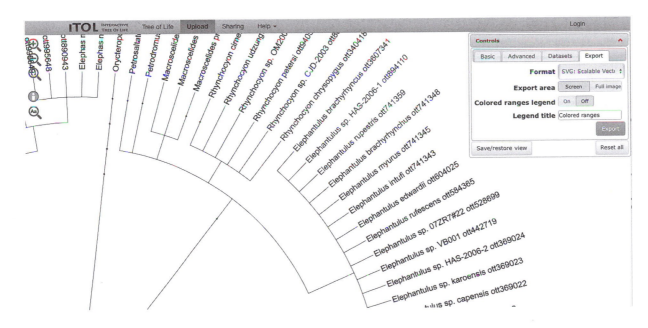

Figure 4.14 A portion of the circular phylogeny plot of mammals from the Interactive Tree of Life website.

5 Keeping a Field Notebook

Time Required

- One 3- to 4-hour lab period, or over the course of several days. Ideally, students will continue to use these methods to develop a complete set of field notes for the course.

Learning Objectives

- Learn how to record observations in a field notebook.
- Understand and apply a standard format for field notes.
- Hone observational skills in the field.
- Understand how to read a topographic map.

Equipment Required

- Field notebook
- Pencils and sharpener
- Topographic map (of the area)
- Field guide to mammals (local or regional guides)
- Other field guides (optional)
- Binoculars (optional)
- GPS device (optional)
- Digital camera (optional)

Background

One of the most basic skills of any mammalogist is the ability to keenly observe the world and then to use those observations to draw conclusions, ask questions, or pose hypotheses. Honing your observational skills is a lifetime pursuit. Obviously, it is much easier to develop good observational skills if you already know something about the species' habitat, behavior, and ecology. At first you will need to rely on other materials such as field guides and identification keys to help you characterize your observations. As you develop your skills you will rely less and less on these tools. Field mammalogists learn a great deal by patient observation and careful note taking. General rules for good field observation include the following:

- Be curious about what you see.
- Slow down, be patient, and watch closely.

- Pay close attention to your surroundings: who (species present), what (behaviors and habitats), when (time and date), where (location, GPS coordinates, or map descriptions), and how (how is the animal interacting with its environment?).
- Be aware of plant communities, microhabitat preferences, interactions between individuals, and any unusual activities.
- Ask questions about your observations, sticking to questions that can be answered by further observations.
- Know your limits—what affects your concentration, what are you likely to overlook.
- Use available field guides, binoculars, GPS devices, cameras, and local experts to enhance your observations.

Why Keep a Field Notebook?

Mammalogists use field notes to record observations, remember the location and details of events, record behaviors, document the flora and fauna observed at a site, and record data for later analysis. Our memories may fade with time, but we can always refer back to our field notes to recall certain events, dates, or measurements. Important discoveries are often made when mammalogists compare recent observations with those taken in the past. For example, comparing the date of first emergence of hibernating ground squirrels with similar dates from many years ago may yield insights into the effects of climate change. Likewise, it is through careful observation and patient recording of behaviors that researchers have been able to decipher mating strategies, courtship behaviors, and many social interactions. It is crucial to write down observations as they occur in the field and not to trust your memory and try to record them later. Field notes also allow researchers to relocate particular sites many months or years later. Many field notes include sketches or maps of the locality and the immediate environment; a map of one of the author's study sites in Madagascar is illustrated in the field notes in Figure 5.1.

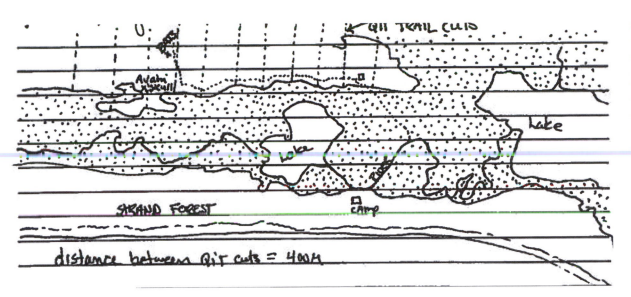

Figure 5.1 A sketch of a field site in southern Madagascar showing the relationship of trap lines to the swamp (stippled) and the base camp.

Each journal entry should include a list of species you encountered that day. It is generally a good idea to separate species by taxonomic group; put all the birds in one list, the mammals in a second listing, and so on, and include scientific names when possible.

One of the most important uses of a set of field notes is to document sightings (or captures) of a species. The detailed records accompanying the sighting in the field notebook provide authenticity of the species location or behaviors; this is particularly important in mammal surveys in which voucher specimens are not collected. Even if a voucher specimen is collected, field notes add critical information about habitat, weather conditions, collection methods used, and many other details that may prove useful in the future.

Elements of a Field Journal

Whether you follow a strict methodology, such as the Grinnell method described below, or simply keep detailed records of field trips, each journal entry should always include certain elements:

- **Date:** The first element of each entry should be the date and the time of day. Dates tell us about seasonal changes as well as placing events in a chronological order. Without a date the journal entry lacks a great deal of its usefulness. Write dates using the day-month-year format. For example, "19 Dec 1989" refers to the 19th of December in the year 1989. Using the familiar US month-day-year format can get you into trouble. For example, 1/4/2010 could be read as January 4, 2010, by an American or as April 1, 2010, by a European. Always using the day-month-year format and spelling out the month will prevent this confusion. Some researchers prefer to place the date at the top of the page, others along the left margin or with the location for each entry.
- **Location and route:** The location and route information should be as specific as possible. Locations should begin with the broadest geographic category (usually country or state), followed by increasingly fine levels of detail. For example, in Figure 5.2 the location is given as "Madagascar, Fivondronana de Tolañaro, Forest along strand, 2.5 km S Manafaify (St. Luce), 24°48′ S, 47°11′ E." This location begins with the country (Madagascar), followed by the district (Tolañaro) and a specific location (2.5 kilometers from the town of Manafiafy). "St. Luce" is given in parentheses because this is the old French name for the town. Whenever possible, give the exact latitude and longitude. Today, handheld GPS devices make it relatively easy to add latitude and longitude. Nevertheless, it is important for any field biologist to know how to extract that same information from a map (batteries die at the worst possible times), and you should acquaint yourself with how to do so. For field sites within the United States, it may be adequate to simply list the state, county or parish, and nearest town, along with the latitude/longitude coordinates. Regardless of where you are, be precise and assume that others may want to use your notes to locate your study site in the future. Ask yourself, can they find my location easily from the information I've provided in my notes?
- **Weather:** Record general weather features such as temperature, precipitation, and cloud cover. A cool, cloudy day may yield few observations, and a moonlit night may yield fewer captures.
- **Habitats:** Record information about the plant community, forest type, and water sources near the site. These may be important when comparing sites.

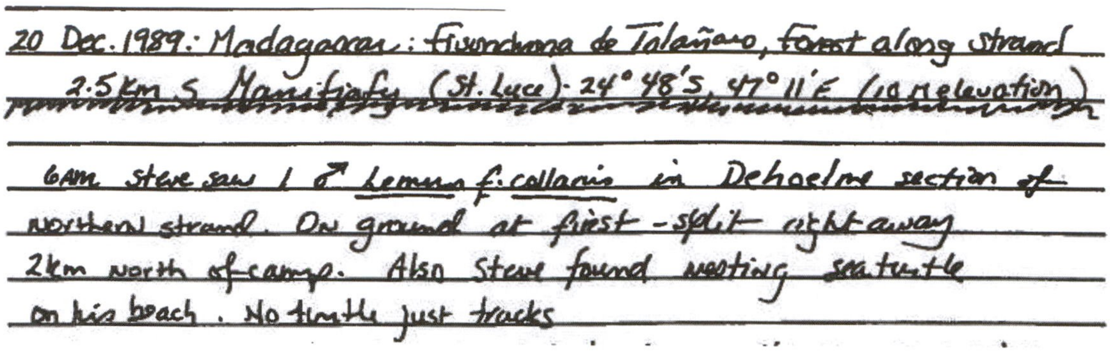

Figure 5.2 A typical entry from the author's field notes for a mammal survey expedition to Madagascar in 1989. Notice that the entry begins with the date and location followed by latitude/longitude and elevation.

> 20 Dec. 1989: Madagascar: Fivondronana de Tolañaro, forest along strand 2.5 km S Manafiafy (St. Luce) 24° 48'S, 47° 11'E (10 m elevation)
>
> 6AM Steve saw 1 ♂ Lemur f. collaris in Dehoelne section of northern strand. On ground at first - split right away 2 km north of camp. Also Steve found nesting sea turtle on his beach. No turtle just tracks

Sketches or hand-drawn maps may be a useful way to illustrate key elements of the habitat, locations of traplines, or den sites within the habitat (see Figure 5.1).

- **Vegetation:** It may be useful to document the dominant plant species, as well as any fruiting/flowering species, as these plants may serve as important food sites for wild mammals and birds.
- **Commentary:** Include any other observations of unusual activity, descriptions of your collecting methods, and the like in short, descriptive sentences. The more detail you can provide the better. Try to anticipate how you will use these data in the future, and attempt to provide any information that may prove useful.

Two-Part Field Notes

Some researchers prefer a two-part process for recording field notes. In this system, you use a small field notebook as the primary record keeper; it travels into the field and is used to record observations. A small, inexpensive flip-top notepad is often used. This primary field notebook is typically small enough to fit into a pocket, leaving hands free for quick access to binoculars and other equipment. Use pencils exclusively for writing field notes. Pencils write in the rain without running and are much easier to use when sketching behavior or maps because they can create shades of light and dark gray.

Later, you will transcribe your field notes into a larger, more formal field journal and annotate them. This secondary field journal—which may be a loose-leaf notebook, a bound book, a soft-cover composition book, or even a word-processed file on the computer is a permanent record. It contains all field observations, along with ideas and predictions derived from those observations. Each entry still contains the essential elements described above for each day: date, location, route, and so on. Entries are typically written in complete, descriptive sentences. Species of interest observed in the area are listed at the end of each entry. Species collection records (with weights and measurements) are usually kept in a separate section of the field journal.

Many mammalogists like to keep an annual field journal, switching to a new one at the beginning of each year or field season. Others prefer a separate journal for each expedition or trip. Professionals always make backup copies (photocopies or digital backups) of their field journals and store them in a separate location from the originals.

Grinnell Method

Joseph Grinnell (1877-1939), former curator of the University of California Museum of Vertebrate Zoology and influential vertebrate biologist, advocated a method that has been widely adopted (with modifications) by mammalogists around the world (for more information on Joseph Grinnell, visit http://mvz.berkeley.edu/Grinnell.html). Grinnell's system uses a tripartite format all integrated into the field journal; the three parts are the journal, the species account, and the catalog:

- **Journal:** In the Grinnell system, the field researcher uses a loose-leaf notebook to record the events and observations of the day as they occur in the field. Each entry begins with the date, time, specific location, and weather conditions, placed in the upper-right corner of the first page of the entry. Every time the researcher moves to a new location, a new page is used with a new heading. Within a location, observations are recorded in chronological order (often with times listed in margins). Grinnell drew a set of vertical and horizontal margins on each page. The vertical margin was approximately 1.5 inches (3 cm) from the left edge and served to provide room for binding the loose pages into a hard cover copy at a later date (Grinnell advocated placing all such bound notes in a suitable museum or library for the use of other scientists). The horizontal margin was approximately 1.5 inches down from the top of the page. In the upper left-hand corner of each page was the observer's name and the year. Only the front of each page was used for writing; the back was often used for sketching maps.

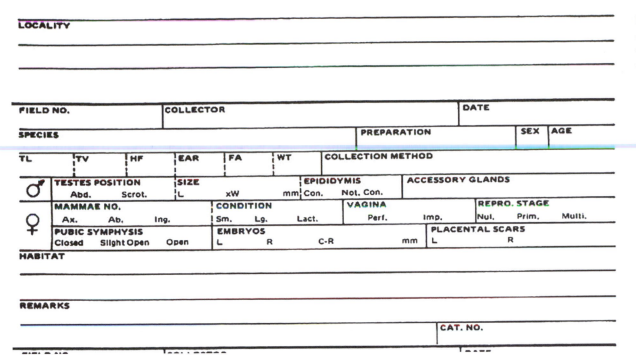

Figure 5.3
A sample of part
of a catalog page
as used by the
University of
Michigan Museum
of Zoology.

- **Species account**: Entries in this section are organized by date with a horizontal line separating each date. Below the horizontal line is a list of the species observed (or at least relevant species). Both common and scientific names are given, where possible, followed by information on the species behavior, ecology, morphology, and so on.
- **Catalog**: These pages are titled "Catalog" in the upper center of the page. Catalog entries list the specimens actually collected at the site and deposited in a museum collection for future reference. These listings are less common today as emphasis has shifted away from the collection of specimens. However, there are times when collection of a voucher specimen is required, or when blood or other tissues are collected non-invasively (e.g., to collect DNA samples).

A sample catalog data sheet appears in Figure 5.3. Because the Grinnell method is time consuming, and requires discipline, a variety of field researchers have simplified this format, retaining Grinnell's basic ideas as a way to document field observations.

Exercise 1: Locality Information Using Topo Maps

You will be given a topo map of a study site by your instructor. Once you are at the site, select a suitable spot for your observations.

1. Open your field notebook and begin by entering the date and location at the top of the entry page. To do this, you will use the following basic format: Day, month, year: country, state, county, township, distance and direction to nearest town (with name of town): latitude, longitude (see instructions below).

2. Locate your study area on the topo map provided, and determine the distance to nearest town or landmark.

3. Locate your exact position on the topo map, and place an X at that spot in pencil on the map.

4. Use your topo map to record the latitude and longitude in your field notebook, and the Universal Transverse Mercator (UTM) coordinates as described in the next section.

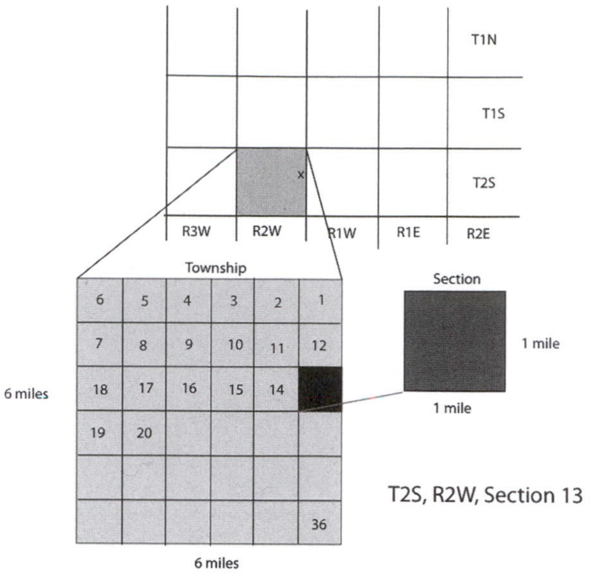

Using Topo Maps

First look at the margins of your topo map, where a number of important features are listed. The title of your map will be in the upper right-hand corner. This corner gives the name of quadrangle and the state where the quadrangle is located (Figure 5.4). It will also give the map series, which indicates how much land area the map covers. A map with a 7.5-minute series covers 7.5 minutes of latitude and 7.5 minutes of longitude.

In 1812, the US government created a standardized system to more accurately define a given US location. This system divides land into 36-square-mile units called townships. Each township has a township and range designation to define its 36-square-mile area. Township is numbered north or south from a selected parallel of latitude called a baseline. Range is numbered west or east

of a selected meridian of longitude called a principle meridian (Figure 5.5).

Not all US topo maps illustrate townships and range numbers. When they are used, townships are subdivided into 36 one-square-mile parcels called sections. Sections are numbered from 1 to 36 for identification. On US Geological Survey topo maps, section, township, and range numbers are printed in red. Section numbers are typically printed in the center of the section on the map (look for a red box with a number in it). Township numbers are printed along the right and left edge of the map (e.g., T.2S and T.3S). Range numbers are printed on the top and bottom edge of the map (e.g., R.1E and R.2E).

Map Scale

The scale on the topo map is found at the bottom center of the map. Scale bars represent a graphic scale with distance in miles, feet, and meters (Figure 5.6). Use the graphic scale to estimate the distance from your site on the map to the nearest major landmark (e.g., town). The ratio scale is listed at the top of the scale bars. For example, 1:24,000 scale means that one inch on the map represents 24,000 inches on the ground (or 2,000 feet).

Latitude/Longitude

To determine the latitude and longitude of your position, locate the numbers running all around the outside of the map (Figure 5.4). These numbers are used to represent two grid systems: the latitude and longitude system and the UTM system. The exact latitude and longitude is given at each corner of your map and at equally spaced intervals between the corners. For the map in Figure 5.4,

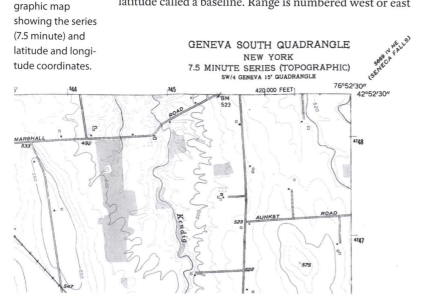

Figure 5.5 Township, range, and section information and sizes for a typical topographic map.

Figure 5.4 The upper-right corner of a typical topographic map showing the series (7.5 minute) and latitude and longitude coordinates.

GENEVA SOUTH QUADRANGLE
NEW YORK
7.5 MINUTE SERIES (TOPOGRAPHIC)
SW/4 GENEVA 15' QUADRANGLE

the latitude and longitude are 76°52′30″ by 42°52′30″. The UTMs are the smaller bold numbers that run along the border of the map (e.g., [47]48). Recall that latitude is distance measured in degrees and minutes north and south of the Equator. The Equator is at 0 degrees, and the poles are at 90 degrees. In the Northern Hemisphere, latitude is always given in degrees north, and in the Southern Hemisphere it is given in degrees south. Longitude is distance measured in degrees east and west of the Prime Meridian, which is 0 degrees longitude. As you go east or west from the Prime Meridian, the longitude increases to 180 degrees, and in the Eastern Hemisphere longitude is given in degrees east.

Degrees are not accurate enough to find a precise location. At best, one degree of latitude and longitude would define a 70-square-mile area. To get finer levels of detail, 1 degree is divided into 60′ (minutes). So, one minute (1′) equals 1.2 miles. If even more accuracy is needed, minutes can be divided into 60 seconds (60″), where one second (1″) equals 0.02 miles. The corner of the topo map in Figure 5.4 reads 42 degrees 52 minutes 30 seconds North latitude and 76 degrees 52 minutes 30 seconds West longitude.

Find your location on the topo map, and run a straight line horizontally out to the right margin of the map and another vertical line to the bottom of the map. Estimate your latitude and longitude from those lines, and add these to your journal heading entry. Longitude tick marks are on the top and bottom edges of the map and latitude tick marks are on the right and left edges. Note that the degrees may be left off (as an abbreviation) and you may only see the minute and/or second designations. The intersection of latitude and longitude lines may be noted by cross-marks (+).

UTM Coordinates

The UTM system divides the surface of the earth into a grid (Figure 5.7). Each cell in the grid is identified by a number across the top called the zone number and a letter down the right side called the zone designator (e.g., New York City is in zone 18T, and Phoenix, Arizona, is in zone 12S). Every spot within a zone can be defined by a coordinate system that uses meters. Your vertical position is defined in terms of meters north, and your horizontal position is given as meters east. They are sometimes referred to as your northing and easting. In Figure 5.4 you can see the northing and easting coordinates on the border of the topo map. They are the small bold black numbers. Along the top left edge of the map the first UTM shown is [3]44 meters. This is actually read as [3]44[000]m. On a regular topo map the dash next to that number would be blue. As you go across toward the right-hand side of the map, the next UTM is [3]45[000] meters. Going from [3]44[000]m to [3]45[000]m means you have gone 1,000 meters. On the map in Figure 5.4 these numbers are abbreviated to [47]47 and [47]48 on the right side and [3]44 and [3]45 across the top.

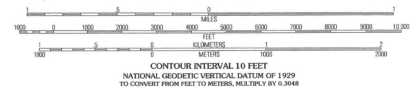

SCALE 1:24 000

CONTOUR INTERVAL 10 FEET
NATIONAL GEODETIC VERTICAL DATUM OF 1929
TO CONVERT FROM FEET TO METERS, MULTIPLY BY 0.3048

Figure 5.6 Scale bars on a typical USGS topographic map.

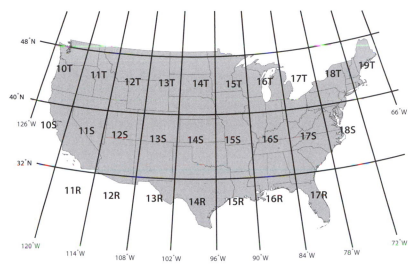

Figure 5.7 A portion of the UTM grid that covers North America.

Suppose you have a UTM position of 18 580647.25 E by 4504637.61 N. The first number in this UTM is 18, which is the zone number, so this map is in zone 18. The next number is 580647.25 E (i.e., your "easting"). This means that on a topo map the position is 647.25 meters to the east of grid line 580. The final number is the "northing" and it tells you how many meters north of a particular grid line the position is. For our example, the position is 4637.61 meters north of grid line 450. (Use Google Earth to find this landmark site.)

To determine the UTM coordinates for your position at your study site, you need to use the map scale. If you are using a 7.5-minute topo map, then it probably has a 1:24,000 scale. This ratio scale means that 1 millimeter (mm) measured on the topo map is equal to 24,000 mm (or 24 meters) on the ground. Using a metric ruler, measure the distance in millimeters from a given UTM grid line to your position (the X you drew on the map) and multiply by 24. This is the number of meters to use in the UTM coordinate for either easting or northing. If you're using a 1:50,000 scale map, then just multiply the number of millimeters by 50. Additional information can be found in Kjellstrom and Kjellstrom Elgin (2009) and Carnes (2007).

Estimate your UTM coordinates, and add them to your journal heading entry. Check your estimates by verifying them with a handheld GPS device. Have your instructor verify that your entry is correct. Move to a new site and repeat the process.

Exercise 2: Taking Field Notes

Find a comfortable spot, sit down, and patiently begin observing and recording notes into your notebook. Be sure to note the current weather, temperature, habitats, and other conditions. Record all plants and animal species you can identify. If you need to, look up the names in field guides (if provided). Spend some time recording the behaviors of one or more focal animals. Record questions that you are curious about. Turn in a copy of your field notes at the end of the session.

6   Livetrapping Small Mammals

Time Required

- At a minimum this lab will require one 3- to 4-hour period to set up the trapping grid, followed by the checking of traps twice a day—in the early morning and early evening—for three consecutive days. The time it will take to identify and process the captured animals will depend on the density of the population and the skill of the students.

Learning Objectives

- Understand the use of small mammal surveys.
- Learn how to set up a standard trapping grid using Sherman live traps.
- Learn how to handle small mammals.
- Practice marking, measuring, and releasing small mammals.
- Use the mark recapture data to estimate population size (see Chapter 9).

Equipment Required

- Sherman live traps (100 traps are typically used)
- Flagging
- Several 50-meter tape measures
- Plastic and cloth bags (to process small mammals)
- Portable battery-operated hair clippers (to mark animals)
- Permanent marking pen (purple or blue)
- Pesola weight scales (60-, 100-, and 300-gram scales are recommended)
- Bait for traps
- Plastic bag filled with cotton or fiberfill
- Small and large rulers for measuring body length
- Backpack
- Respirator masks
- 5–6 pairs of latex gloves (to handle small mammals)
- Leather gloves (to handle larger mammals)
- Plastic bags (to process small mammals)
- Cloth bags or pillowcases (to process larger mammals)
- Field journal
- Pencils and pencil sharpener
- Extra Sherman traps
- Drinking water

Background

Small mammals are generally defined as mammals weighing less than 5 kilograms (kg) as adults (Merritt, 2010). This category usually includes rodents, bats, marsupials, shrews, moles, hedgehogs, elephant shrews, and a few species of small carnivores and primates (Figure 6.1). Of the more than 6,400 species of mammals, nearly 90 percent (i.e., 5,760 species) are considered small mammals (Wilson and Reeder, 2005). Understanding small mammal ecology and behavior requires a different approach than would be taken for large, diurnal mammals. Small mammals are often secretive; many species are nocturnal, semi-fossorial, or otherwise difficult to observe. Obviously, many questions can be answered only by collecting detailed data on small mammal populations over long periods of time.

Examples of data requiring long-term study over multiple seasons or years include

- home range and territory sizes
- population fluctuations over time
- effects of changing land use on small mammal communities (e.g., increasing urbanization)
- colonization patterns
- predator/prey relationships
- effects of natural disasters (e.g., floods, fires)

Other questions can be answered by shorter, more concentrated study. Data from such studies might include

- species diversity in a region (e.g., rapid assessment programs)
- relative abundance or density of species
- habitat preferences
- reproductive condition in a given season
- diet

To collect data for both long- and short-term studies, it is often necessary to capture and handle mammals. Animals must be captured, for example, in order to

Figure 6.1 A mouse lemur (*Microcebus rufus*, Primate) captured by the author in Madagascar.

attach radio transmitters, ear or passive integrated transponders (PIT) tags, or to collect tissue or DNA samples. One particularly useful and frequently employed method is to collect data from livetrapping studies. In this lab exercise, you will set up a standard trapping grid using Sherman live traps for small mammals, collect mark and recapture data, and estimate population size from those data.

The general procedures for livetrapping are as follows:

- Set out a transect or grid array of uniquely marked live traps.
- Bait and open each trap on a specific schedule.
- Check each trap on a schedule that minimizes stress to the animals.
- Record the species, trap number, sex, and any other data deemed necessary for each animal captured.

- Mark and release each animal at the trap site where it was collected.
- Re-bait and set traps again for the next session.

Your instructor will provide additional details about how to bait and set the traps and how to mark animals that are captured. There are risks associated with capturing and handling live mammals. It is recommended that only professionals supervise small mammal trapping and handle the mammals to avoid harm to the inexperienced researcher and to the captured mammals.

Your instructor should review the guidelines on humane capture and handling with you before the study begins. Remember, "Investigators conducting research requiring live capture of mammals assume the responsibility for using humane methods that respect target and non-target species in the habitats involved" (Sikes et al., 2016, p. 673).

Live Traps

A Sherman live trap is a small aluminum folding trap that unfolds into a rectangular box. There are two sizes, but in most cases the larger 3″ × 3.5″ × 9″ size works best for many small mammals. The advantage of folding traps is that they are easier to transport and clean. Information on Sherman traps can be found at http://www.sherman traps.com/.

Traps are unfolded in the field and baited with a mixture of rolled oats and peanut butter (or other comparable food items that can form a thick paste). Place a small ball of bait directly behind the treadle (spring-plate inside trap). Some professional mammalogists prefer to wrap the bait into a small bundle using cheesecloth to avoid having to clean the traps as frequently. To capture shrews and other insectivores, or small carnivores, it may be necessary to use a small piece of sardine or ground beef instead of oats and peanut butter. Once you have placed the bait inside, push the trap door down to the floor until it catches behind the small treadle plate. You can adjust the holding latch carefully using your fingers. Attempt to set the treadle so that it just holds the door in place.

Exercise 1: Setting Up a Livetrapping Grid

The first step is to choose a suitable study site. Livetrapping grids are generally preferred for mark-recapture studies. The size of the grid area will depend on the number of traps available and research question being asked. Generally, traps are spaced at 10-meter intervals in 10 rows of 10 traps each (requiring 100 traps), but smaller grids, which have the advantage of requiring fewer traps, may be adequate (Figure 6.2). Whatever the size of the grid, each trap row should contain an equal number of traps, and trap rows should be spaced 10 meters apart.

IMPORTANT: *Before conducting any livetrapping study, be sure to obtain the correct permits for your state (or region). Collecting permits can usually be acquired from the state or regional conservation department. In addition, it is important to obtain approval from your institution's Institutional Animal Care and Use Committee (IACUC), even for work that occurs off-campus; consult with your institution's IACUC for advice. Finally, it is important to follow the American Society of Mammalogists guidelines for field work with mammals (Sikes et al., 2016). These guidelines can be found on the web at http://www.mammalsociety.org/committees/animal-care -and-use#tab3.*

1. Begin by marking off one side of the grid with a 100-meter tape (or by pacing it off).

2. Every 10 meters (or paces) along the length, place a flag and label the flags A through J.

3. At each lettered flag, measure a second 100-meter line perpendicular to the original line. There will be 10 such lines (one for each lettered flag) all on the same side of the original line, forming a grid.

4. Place a flag at each 10-meter interval along these perpendicular lines and label each flag with the letter of the row and a number representing the number of the 10-meter interval (Figure 6.2). Each flag, marked with a unique letter-number combination (e.g., A1, C7, B10), records the future position of a trap on the grid.

5. Check the grid to make sure that the parallel lines do not converge or diverge too much. For more permanent grids in open habitats, it may be necessary to use surveying equipment.

6. Once the grid is established and marked, place one Sherman live trap near each flag in the grid. Placing the traps so as to maximize the probability of capture is an art. Experienced mammalogists look for places small mammals are likely to travel, such as along downed logs, near stumps, and within grassy clumps (Figure 6.3). In an old field, traps should not be placed on top of the grass, but within the dead grass layer close to the ground. Try and position each trap within 1 meter of the flag. In tropical regions, traps may be placed above ground along horizontal branches as well as on the ground. Be sure to record the position of each trap that is placed above the ground in your field notes. Each trap should have a small piece of painter's tape (e.g., blue tape) with the letter-number code for that trap placed on the top of the trap (the tape can be easily removed during cleaning).

If the grid site or transects are to be used in future months or years, permanently mark each trap location or record the coordinates of the corners of the grid with a GPS unit so that they can be easily found again.

Figure 6.2 A typical trap grid of 100 trap stations (squares) laid out as a 10-by-10 grid. Note that each trap station has a unique letter and number code (e.g., A1 or H7).

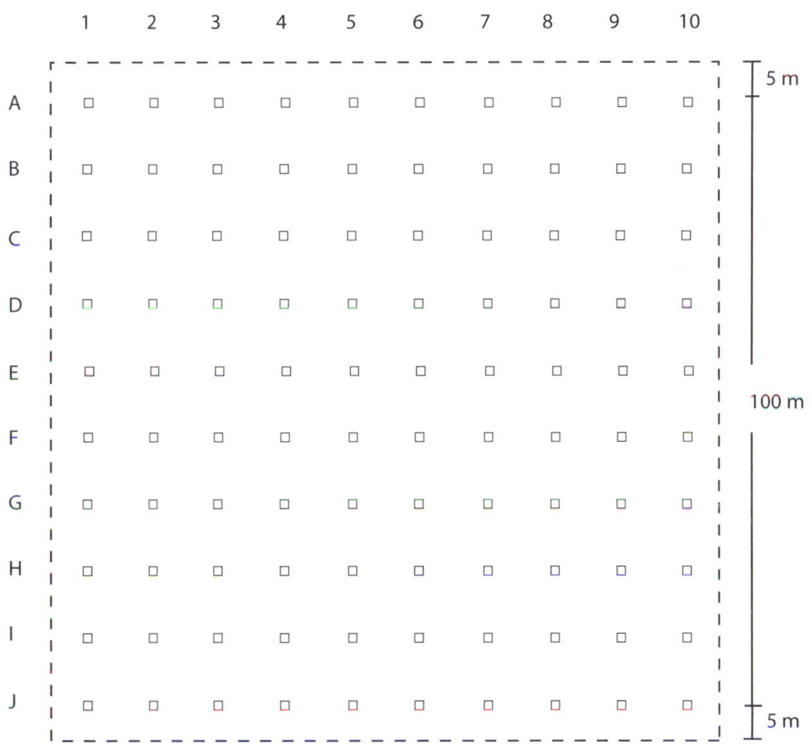

Figure 6.3 A Sherman folding trap with the trap door open and set adjacent to a log.

Exercise 2: Checking Traps and Collecting Capture Data

For a general mark-recapture survey, traps should be set for at least three nights. Check traps twice during the day—once in the early morning and again in the evening. This schedule reduces mortality and allows both nocturnal and diurnal animals to be captured. To avoid accidental trap deaths, never set traps in direct sunlight or when cold snaps are forecast. If poor weather threatens, add a small wad of fiberfill to the back of each trap to provide insulation until the traps are checked. After you have set the traps, check to see that each trap's door is in the open position.

1. Check the traps at dawn the following morning, shortly after the sun rises. Waiting longer can lead to unnecessary animal death.

2. Carefully evaluate each trap in the row. Check to see if the door is still open; if not, note the trap's position on the grid in your field notes. To check a closed trap, carefully lift the trap so that the front door is facing upward (do not shake the trap). To avoid inhaling urine or feces, keep the trap away from your face. Slowly push open the trap door just enough to see if an animal is inside.

3. To remove the small mammal from the trap, hold the trap with the trap door facing upward, and place a large plastic or cloth bag around the outside corners of the trap and over the door.

4. Slowly turn the trap and bag upside down (so the animal is on the "ceiling"), and gently push open the door (with the bag still sealed around the door).

5. Slowly tilt the trap until the animal slides into the bag. Shake the trap lightly if needed.

6. Close the bag and weigh the bag and animal using a Pesola spring scale (Figure 6.4). Always weigh the empty bag when done, and subtract the bag weight.

7. Identify the mammal to species.

8. While the animal is in the bag, gently move it toward the bottom of the bag and take a measurement

Figure 6.4 Pesola spring scale used in small mammal field research.

of total body length and tail length using a small metric ruler.

9. While the animal is still in the bag, grab it by the back of the neck (through the bag), and invert the bag to expose the animal.

10. While holding it securely, mark it, determine the sex (if possible), and use a small plastic ruler to take any remaining measurements. Work quickly and methodically to avoid undue stress to the animal.

11. When marking is complete, place the bag close to the ground, open it, and wait for the animal to exit.

12. All traps should be re-baited (if needed), opened again, and replaced in their original position.

Handling Captured Mammals

Depending on the size of the animal and the procedures required to mark or tag it, captured animals might require chemical immobilization (e.g., anesthesia) for longer handling periods. In many cases an experienced field mammalogist can remove, mark, measure, sex, and release an animal in a minute or less. However, if a radio collar is being attached, anesthesia is generally used. It is important to remember that immobilization may in some cases cause more stress than working efficiently without immobilization. The American Society of Mammalogists recommends that chemical immobilization be used on a case-by-case basis when animals are sub-

ject to "more than momentary or slight pain or distress" (http://grants.nih.gov/grants/olaw/references/phspol .htm). The choice of anesthetic or analgesic and its dose and route of administration should be made in consultation with a wildlife veterinarian (National Research Council, 1996).

Holding animals without causing harm to the animal (or potential bites and scratches to the researcher) takes experience. In general, handling should be kept to a minimum. Pick up small mammals by the scruff of the neck, using thumb and first finger. The hand is rotated so the animal's belly faces the researcher and the animal's tail is grasped with the pinkie finger. Animals heavier than 200 g require a firmer hold on the body

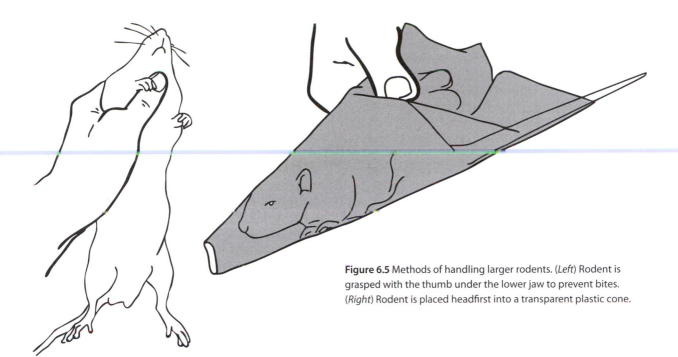

Figure 6.5 Methods of handling larger rodents. (*Left*) Rodent is grasped with the thumb under the lower jaw to prevent bites. (*Right*) Rodent is placed headfirst into a transparent plastic cone.

and isolation of the head and jaws (or anesthesia) (Figure 6.5).

CAUTION: *Never hold a wild small mammal by the tail; it can cause damage to tail vertebrae and even tail loss in some tropical species. Species that are aggressive or potential disease reservoirs (e.g.,* Rattus norvegicus*) should not be handled without gloves. In such cases, researchers often use a special plastic cone to immobilize the animal (Figure 6.5). These heavy plastic cones are transparent and have a small opening at the tip so the animal can breathe. The rodent is placed in the cone headfirst to immobilize it. With the animal so restrained, it is possible to weigh it, collect tail blood, or inject the animal without getting bitten.*

Marking Mammals

Techniques for marking individual animals should be non-invasive, allow the researcher to uniquely mark many individuals, allow identification over a long period, and allow marking by one person.

There are many methods for marking small mammals. Traditional methods include toe clipping or attaching tiny numbered ear tags. Newer, less-invasive methods include inserting PIT tags underneath the skin, or clipping or dying patches of hair. Each method has its advantages and drawbacks. For example, PIT tags can cause injury at the injection site and are expensive.

Two methods that have been successful for shorter-term studies are hair clipping and dying. Hair clipping involves using battery-powered grooming clippers to cut unique patterns of lines into the fur on the animal's flanks. To ease future identification of that individual,

the design and position of the marking on each animal are recorded on the data sheet or a photograph is taken. Alternatively, Sharpie permanent markers can be used to write a number or letter on the animal's belly. Use blue or purple color markers as other colors are harder to read or may be confused with blood or dirt. Remember, fur must be dry or the ink will smear.

Sexing and Aging Small Mammals

Sexing small mammals can be difficult. Usually the external genitalia of the male can be distinguished from those of the female by their position relative to the anus and by size. Some groups (e.g., shrews and some marsupials) are difficult to sex without dissection, because the testes remain in the body even when the males are sexually active. Furthermore, in mammals both sexes have nipples, although they are less prominent in males. Adult male rodents will have testes in the scrotal sacs during the breeding season, but outside of the breeding season the testes are in the abdomen (recrudescence), making sexing difficult during the nonbreeding season.

The best method of sexing rodents in the field is to note the relative distance between the anus and the penis (males) and vaginal opening (females). As illustrated in Figure 6.6, the distance is greater in males than in females.

Once the sex has been established, it is often useful to record the reproductive status of the individual (Gurnell and Flowerdew, 2006). As previously mentioned, males are considered reproductively active if their testes have descended into the scrotal sac. Reproductively active adult females are distinguished by

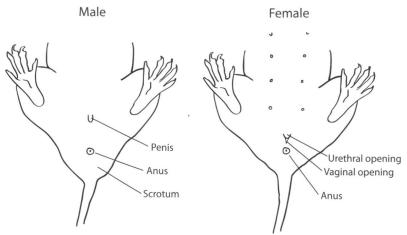

Male   Female

Penis
Anus
Scrotum

Urethral opening
Vaginal opening
Anus

Figure 6.6
Diagrams of the ventral abdomens of a male (*left*) and female (*right*) rodent illustrating the relative positions of the genitalia and anus.

It is not possible to age most live small mammals exactly. Therefore, they are divided into age categories: neonate (or young), juvenile, and adult. Young are smaller, lack fur or have softer fur, and have proportionately large heads. Juveniles have the same proportions as adults but tend to have gray pelage and are smaller than adults. Adults are individuals capable of reproducing.

Measuring Small Mammals

Animals should be weighed and measured as quickly as possible. Weights should be made to the nearest half-gram (greater accuracy is not needed and, considering the scale on spring balances, unlikely to be real). The standard measurements are

- presence of perforate vagina (a small opening is present)
- presence of seminal plug in vagina (males produce seminal plugs during copulation, but these plugs are transitory and are not produced by all species)
- distended abdomen (gentle palpation of the abdomen may indicate a pregnant female)
- presence of milk in nipples (lactation)
- presence of halos (bald patches around the nipples from sucking action)

- total length (TL) from the distal tip of the last tail vertebra to the tip of the nose
- tail length (T) from the distal tip of the last tail vertebra to the vertex where the tail meets the body (bend the tail 90 degrees to find the vertex)
- hind foot length (HF) from the back edge of the heel to the tip of the longest digit including the claw
- greatest ear length (E) from the notch at the base of the ear cavity to the tip of the pinna
- body weight (W), measured (in grams) using a spring scale (e.g., Pesola) to the nearest half-gram

Exercise 3: Data Collection and Analysis

Record all of your recapture data on the data sheet provided as well as in your field notebook. Combine all of the data from all students and trapping sessions into a master data table for the study.

   1. Using the data collected for your grid trapping period, create a summary table of the total number of individuals captured for each species and the trap success (percentage of each species captured per trap night). Use the format shown in sample Table 6.1.

   2. Calculate the total number of trap nights (effort). (Trap nights = number of traps open per 24-hour period × number of 24-hour periods.)

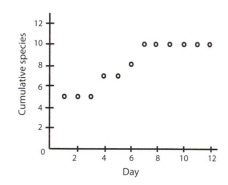

Figure 6.7 Graph of the cumulative captures for all species of small mammals over a 12-day period.

Table 6.1 Hypothetical summary table for a grid trapping session

Scientific name	Common name	Number captured	Percentage captured
Peromyscus maniculatus	Deer mouse	296	64.9
Mictotus pennsylvanicus	Meadow vole	142	31.1
Sorex cinereus	Masked shrew	12	2.6
Blarina bervicauda	Short-tailed shrew	6	1.3
Totals		456	100.0

3. Calculate the cumulative number of species captured over the entire trapping period (Figure 6.7).

Using the data from your trapping session, estimate the density of each small mammal species population on the grid. First, calculate the number of unique individuals of each species on the grid. Divide the number of each species by the total area of the grid to get an esti-mate of density of small mammals on the study site. Add a boundary strip to the area of your grid to account for the fact that traps on the edge of the grid actually sample areas partially outside the grid. For example, if the grid was 100 meters by 100 meters with 10 meters between traps, then you would add a 5-meter (half the distance between traps) strip around the perimeter, making the actual area sampled 110 meters by 110 meters (Figure 6.2).

Appendix

Table 6.2 Data table for a small mammal livetrapping session

Date: _____ Observer: _____ Location: _____

Start time: _____ End time: _____ Temp (°C) _____ Weather _____

Trap number	Species code	Fate	Age class	Sex	Reproductive condition	Recapture	Mark #	Total weight (g)	Bag weight (g)	Animal weight (g)	Total length (mm)	Hindfoot length (mm)	Ear length (mm)	Tail length (mm)	Comments

Notes: Fate: N = new, R = recapture, D = dead. Age class: A = adult, J = juvenile. Sex: M = male, F = female. Reproductive condition: 1 = nonbreeding, 2 = pregnant, 3 = lactating, 4 = testes enlarged, 5 = unknown.

7   Specimen Preparation

Time Required
- The preparation of a study skin may be completed in a typical 3- to 4-hour lab period.

Learning Objectives
- Understand the purpose of museum collections.
- Understand the different methods for collecting data to accompany specimens.
- Learn how to prepare a fluid specimen.
- Learn how to prepare a skeleton and skull.
- Learn how to prepare a museum study skin.

Equipment Required
- Dissection equipment
- Data sheets
- Metric ruler
- Pesola scales (various sizes)
- Specimen tags
- Cotton batting
- Cornmeal or fine sawdust
- Straight wire (12- to 18-inch lengths)
- Wire cutter
- Large syringe with 18-gauge needle
- 10% neutral buffered formalin
- Large glass or plastic containers
- Cheesecloth

Why Collect Specimens?

Collecting mammal specimens is essential for proper scientific documentation of ecological change, conservation, and many other scientific studies. Researchers from natural history museums and research institutions worldwide agree that there are no alternatives to collecting specimens that are reliable for identifying species or documenting ecological changes within populations or ecological communities (Rocha et al., 2004; Suarez and Tsutsui, 2004). They argue that properly documented and prepared specimens are essential for wildlife managers and conservation biologists to make informed decisions. Museum natural history collections continue to play a vital role in our understanding of the spread of diseases,

the consequences of invasive species, and changes in geographic distributions accompanying climate change.

An unknown respiratory syndrome appeared in the southwestern United States in 1993 and resulted in the deaths of several people in the Four Corners region. The disease turned out to be caused by a hantavirus found in deer mice (*Peromyscus*; Yates et al., 2002). Genetic analysis of museum specimens revealed that hantavirus was present in *Peromyscus* long before the 1993 outbreak. These museum specimens were critical in establishing that hantavirus only infected humans after rodent populations soared following the massive 1992 El Niño event.

Museum specimens played a critical role in establishing pre-industrial levels of several environmental contaminants including mercury, DDT, and more recently the herbicide atrazine (Berg et al., 1966; Hayes et al., 2002). Museum collections are also being used to document the spread of invasive species and to evaluate their impact on native species (Ward, 2007). Researchers are using museum specimens as baseline data for documenting the effects of climate change on the distribution and breeding ecology of many species. There can be no doubt that museum specimens allow scientists to understand past events and glimpse the future. That said, museum specimens are virtually useless without proper documentation.

Documenting Specimens

When a specimen is collected in the field, mammalogists typically give it a field number and record the data in their field notebook (see Chapter 5). Later, when the specimen is accessioned into a museum collection, it is given a new museum accession number. Copies of all relevant field notes, field catalogs, photographs, and other useful information are given to the museum so that a specimen's accession number can be tied back to the original field notes. Most mammalogists carry a small (5 × 7-inch) loose-leaf binder containing archival-quality paper divided into two sections: a field journal and a field catalog for specimens collected and prepared for mu-

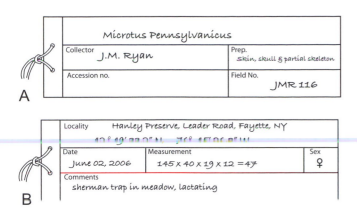

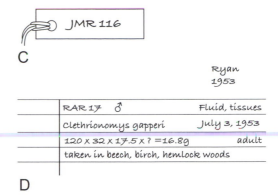

Figure 7.1 Examples of specimen tags: (**A**) front of a typical specimen tag, (**B**) back of the tag, (**C**) a skull tag, and (**D**) a sample entry in a field catalog with no ear length recorded. Note that *Clethrionomys gapperi* is now called *Myodes gapperi*.

seum collections. The reason for a loose-leaf binder rather than a bound journal is that additional pages can be added when needed. The margins of the paper are large enough to allow for binding, so that the notes can later be bound and archived in a museum. This journal is similar to a daily diary of events. It details how many traps were set in each transect, locations of transects, summary records of captures, and interesting notes about habitat or ecology of the specimens captured. The catalog contains specific information that will accompany each specimen collected and preserved for later accession into a museum. All entries are made in waterproof ink or in pencil.

Journal entries are discussed in Chapter 5. Catalog entries for each specimen prepared as a museum specimen are standardized. In some cases, museums provide printed catalog pages to collectors (Figure 5.3). In most cases, collectors create their own catalog by writing their last name at the top of the page along with the date (write out the name of the month to avoid confusion). Each specimen is given a field number. Many mammalogists use their three initials followed by a unique consecutive number (e.g., JMR 116). The field number is followed by the specimen's sex, indicated by the symbols ♂ and ♀, and finally by the type of preparation (skin, skeleton, fluid). On the line below, enter the species name (genus and species) in pencil and the date. If you are not sure of the species you can enter "cf" after the genus name but in front of the species name. If it is later changed, the pencil can be erased and the correct name entered. On the next line down, enter the standard measurements and reproductive condition (see details below). The final line is reserved for comments about the capture location, trap type, or other important ecological information. After the data are entered into the catalog, a specimen tag with the field number is attached to the specimen (Figure 7.1).

It is critical that all mammalogists use the same methods for recording body weights and measurements. Measurements such as tail length and hind foot length are often used to separate species that otherwise look very similar. Imagine how much variation would be added to hind foot length if some mammalogists chose to include the toenails and others did not. To accurately record these measurements, scientists use metric units (millimeters and grams) and always record the measurements in a standard sequence: total length, tail length, hind foot length, ear length, forearm length (for bats only), and weight. Each measurement is separated by "×" (times symbol), and length measurements are separated from weight by "=" (equals sign) (Figure 7.1). If the tail is broken off, it is indicated by a "?" and a note is made in the field catalog.

Exercise 1: Taking Standard Measurements

Practice recording standard measurements on specimens provided by your instructor. For each specimen, fill out a specimen tag and a catalog entry.

- Record your field number, the species name, and the date on the specimen tag and in your catalog.
- Determine the sex of the specimen (refer to Chapter 6).
- Record the linear measurements with a millimeter ruler that is trimmed so that it begins at the zero mark. Record each measurement as follows:
- Total length: the distance from the tip of the nose to the last tail vertebra (do not add the length of

tail tuft hairs). Usually the ruler is placed on a flat surface, and the specimen is placed on top (gentle pressure may be applied to get the specimen to lie flat).
- Tail length: the total length of all caudal vertebrae (Figure 7.2). This is measured by placing the animal on its belly and forming a 90-degree angle between the body and the tail. The zero end of the ruler is positioned at the base of the tail and the rest of the tail is held flat against the ruler. Record the length to the nearest millimeter. You may want to do this several times and take the average.

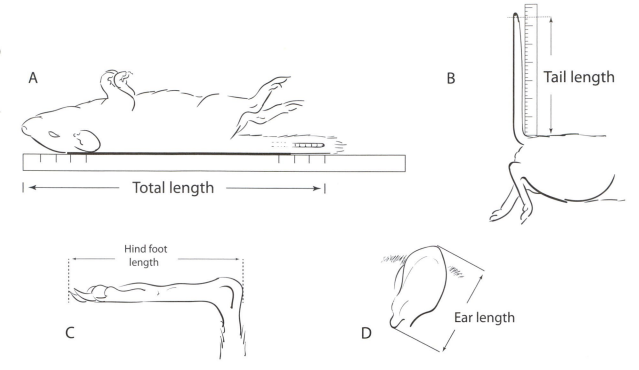

Figure 7.2 Examples of how to measure mammalian specimens: (**A**) total body length, (**B**) tail length, (**C**) hind foot length, and (**D**) ear length.

A

Total length

B Tail length

Hind foot length

C

D Ear length

- Hind foot length: the distance from the tip of the heel to the tip of the claw on the longest toe (Figure 7.2).
- Ear length: the distance from the notch at the base of the ear to the tip of the pinna (Figure 7.2).
- Forearm length (bats only): the distance from the tip of the elbow to the wrist with the wing folded.

- Body weight: the weight (mass) of the unskinned animal recorded in grams. Use a spring scale (e.g., Pesola) of the smallest size available (e.g., 10 g, 30 g, 100 g maximum capacity).

Check with your instructor to see if your measurements match those already recorded for that specimen.

Recording Reproductive Data

Notes on the reproductive condition of collected specimens are important to establish breeding season length, litter size, and age of sexual maturity for a given species. Thus, it is important never to guess. For males, record the position and size of the testes. The position is either abdominal or scrotal, and the size is measured as length and width in millimeters (for a dissected testis). If possible, note whether the cauda epididymis tubules are visible on the testis as these tubules are typically visible only when they are storing sperm.

A female's reproductive status is a bit more complicated. Breeding females can be pregnant, lactating, or both, or not in reproductive season. The condition of the vaginal opening is a good indicator of reproductive status in many mammals, so first note whether the vagina is imperforate (sealed by a membrane) or perforate (open). If perforate, then it is probably the breeding season. Second, determine the condition of the mammary glands. Lactating females usually have somewhat swollen teats, and milk can often be squeezed from the teat.

Pregnancy is determined if the female has an obviously swollen abdomen and if the dissected uterus contains developing embryos. In dissected specimens, count the number of embryos from the top of the uterine horn to the base on each side. For example, the notation "6; 2L, 4R" denotes a female with six embryos—two on the left side and four on the right. If possible, measure the crown-rump length of each embryo. Note if there are placental scars or indications that an embryo has been resorbed. Placental scars are yellow or black spots on the uterus where a placenta was attached. Placental scars can last through more than one breeding season so they are not reliable for determining litter size. They are, however, useful for determining whether the female has bred in the past. If visible, count the placental scars on each side and record the number in your field notes. Females with no placental scars or embryos are referred to as nulliparous. Females with embryos or one set of scars are primiparous, and those with embryos and one or more sets of scars are called multiparous.

Along with measurements and sex, it is best practice to record habitat, trap position, and other relevant information. Habitat descriptions include dominant plants (hemlock forest, or sagebrush scrub), elevation in meters, and any relevant ecological information about the area (recently burned or seasonally flooded). This is especially important for rare species.

The method of capture is also important. Describe how the animal was acquired. Was it livetrapped on the ground, livetrapped on a tree branch, netted over a river, shot, or found dead? Record where the trap was placed and the time of capture, if known. Record any behavioral observations in your journal. These may include observations of roosts, dens, feeding, and weather conditions.

Preparing Museum Specimens

Museum specimens include study skins, partial or complete skeletons, skulls, fluid-preserved specimens, and tissues. The type of specimen depends on the objectives of the study. Study skins are useful for studies of geographic variation or pelage color, whereas fluid specimens are appropriate for anatomical studies or histology.

Preparing small mammals in fluid is a two-step process: initial fixation and long-term storage. After a dead specimen has been measured and assigned a field number, a field tag is tied securely to the ankle of the hind foot. Use paper labels that will not disintegrate in the preservative and waterproof ink or a pencil to indicate the field number and sex on both sides of the tag. Lay the mammal on its back, insert the needle of a syringe filled with formalin into the abdomen, and slowly fill the body cavity with formalin. Do not overfill—the body should not be distended. For larger animals, inject formalin into large muscles as well. Finally, prop the mouth open with a wad of cotton, wrap the specimen in cheesecloth, and immerse it in a container of 10% neutral buffered formalin. The mouth is fixed in an open position to allow easy examination of teeth. Bats are fixed with wings folded. Be careful not to crowd the specimens so that they become fixed in unnatural positions. Complete fixation is achieved in 24–48 hours for many small mammals. Specimens can stay in formalin fixative for up to two months. After that period, they should be thoroughly rinsed in water and transferred to 65 to 70% ethanol or 45 to 60% isopropyl alcohol for long-term storage.

Ten percent neutral buffered formalin is a solution of formaldehyde. Commercial formalin is actually 37–40% formaldehyde, but it is usually treated as 100% formalin. Thus, to make a 10% formalin solution, mix one part 37–40% formalin with nine parts water. It is important to neutralize the acidity of the formalin so that it does not decalcify bones and teeth. For best results, mix 4 g of monobasic sodium phosphate monohydrate ($NaH_2PO_4H_2O$) and 6.5 g of dibasic sodium phosphate anhydrate (Na_2HPO_4) to each liter of 10% formalin. Alternatively, add a teaspoon of borax (sodium tetraborate) to each gallon of 10% formalin.

Traditional museum study skins are filled with cotton to approximate the natural shape of the mammal. The lower leg bones are ordinarily left in the skin. Therefore, a partial skeleton and skull can also be prepared along with the study skin. Large mammals are skinned and the skins are tanned (these procedures are not covered).

Exercise 2: Preparing a Museum Study Skin

Begin with a recently euthanized or thawed frozen specimen. Record all the measurements and field notes as described above. Make a midline incision in the skin over the belly extending from in front of the anus to the side of the genitalia and anterior to about the level of the last rib (Figure 7.3). The incision should not penetrate the abdominal cavity. Gently separate the skin from the underlying muscle to either side of the incision. To keep the skin clean and dry, add some cornmeal, borax, or fine sawdust to the opening to absorb blood and body fluids. Continue to work the skin free of the body wall in the vicinity of the incision and toward one hind leg. Grab the hind foot and push the knee forward towards the incision. Carefully work the skin off the leg down to the ankle. With scissors, cut the hind leg at the knee joint (Figure 7.4). Remove any large chunks of muscle still attached to the lower leg. Repeat this process with the other hind leg.

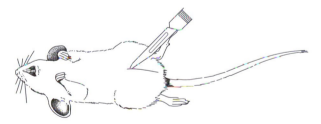

Figure 7.3 The first step in preparing a museum study skin is to make a midline incision in the skin over the abdomen.

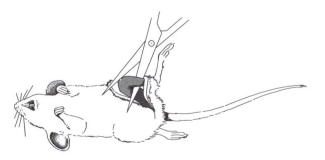

Figure 7.4 Cut the leg below the knee to leave the foot attached to the skin.

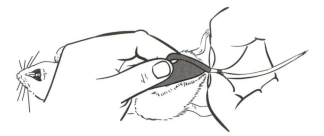

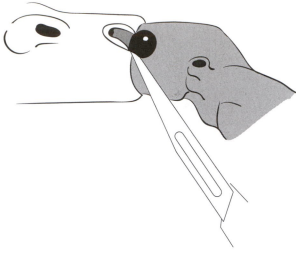

Figure 7.5 Sliding the tail vertebrae out of the tail sheath. Note the position of the fingernails just in front of the tail sheath. The carcass is shown in gray.

Figure 7.7 Cutting the membrane over the eyes and the corner of the eyelid.

When both hind legs are cut, work the skin free down to the base of the tail. Carefully skin around the anus and scent glands. Grasp the skin at the base of the tail with the fingernails of your thumb and index finger placed just in front of the skin (Figure 7.5). Gently slip the tail vertebrae out of the tail sheath by pulling the vertebrae at the base of the tail away from the hand holding the tail skin. This leaves the tail skin intact as an empty tube still attached to the rest of the animal's skin.

With the tail free, continue to peel the skin forward to the region of the front legs. Apply cornmeal or sawdust as necessary to keep the skin from becoming sticky. Do not pull the skin; instead use one hand to gently push the skin away from the body and a scalpel to sever any connective tissue. Carefully remove the skin from the front legs down to the wrist. Using scissors or a scalpel, cut the front legs below the shoulder joint. Peel the skin over the chest area forward to the base of the skull.

The most difficult stage is removing the skin from the head without damaging the ears, eyelids, lips, and skull. Use a sharp scalpel to carefully work the skin over the head to the base of the ears (Figure 7.6). Observe the cartilage at the base of each ear. Using the scalpel cut the cartilage close to the base of the ear. Continue to peel the skin over the head until the posterior margin of the eyes are exposed. With the skin held away from the head, cut the membrane that covers the eyes (Figure 7.7). The skin should be still attached in the eye region at the front corner of the eyelid. Carefully cut this attachment but do not cut into the eyelid. Continue working the skin forward to the lips. Carefully cut the connective tissue at-

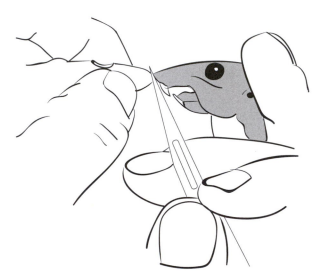

Figure 7.8 Cutting the nasal cartilage in front of the nasal bones to free the skin from the carcass.

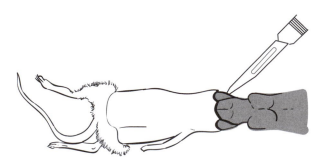

Figure 7.6 The specimen with the skin peeled back to the base of the ears.

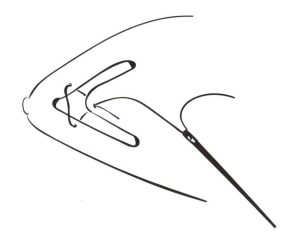

Figure 7.9 A simple triangular stitch is used to sew the lips together.

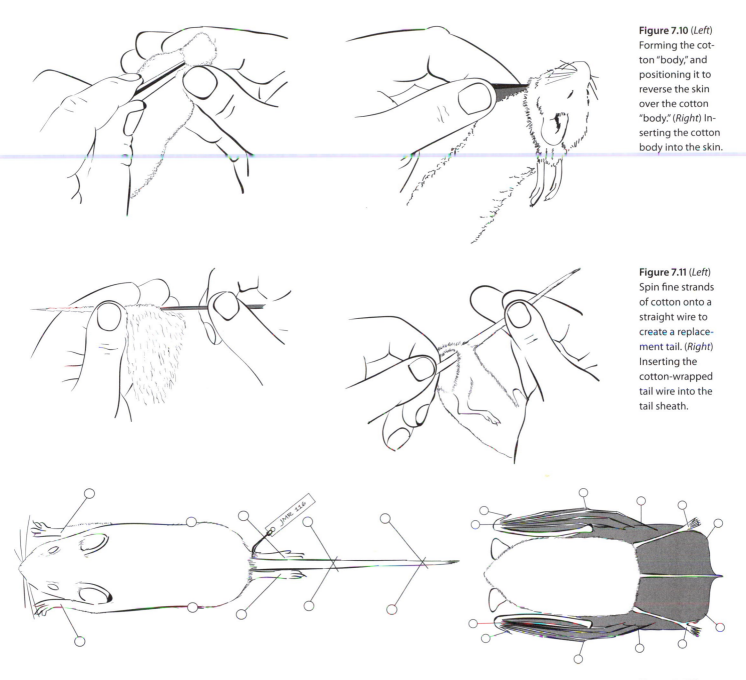

Figure 7.10 (*Left*) Forming the cotton "body," and positioning it to reverse the skin over the cotton "body." (*Right*) Inserting the cotton body into the skin.

Figure 7.11 (*Left*) Spin fine strands of cotton onto a straight wire to create a replacement tail. (*Right*) Inserting the cotton-wrapped tail wire into the tail sheath.

JMR 116

Figure 7.12 The proper placement of pins for drying (*left*) rodents and (*right*) bats.

taching the lips to the skull. Finally, peel the skin forward to the tip of the nose. Cut the nasal cartilage in front of the nasal bones of the skull (Figure 7.8).

The skin should now be free of the carcass. Put the carcass aside for now and remove any clumps of fat attached to the skin. Rub some magnesium carbonate powder or sawdust onto the skin to help accelerate drying (if available). With the skin still inside out, sew the lips together using a triangular stitch (Figure 7.9).

Prepare a cotton "body" to go inside the skin using a single piece of cotton batting. This is best accomplished by rolling the cotton into a cylinder slightly longer and thicker than the body of the mammal (Figure 7.10). Grasping one end with a pair of forceps or hemostat, form a cone shape; this will be the new head. Place the

tip of the cone of cotton against the inverted nose of the skin and reverse the skin over the cotton and forceps. Before removing the forceps, adjust the mouth, ears, and shape of the head and continue reversing the skin over the cotton "body" (Figure 7.10). Trim the cotton with scissors to fit it inside the skin properly.

To replace the tail vertebrae, select a straight wire, trim it to about 2 inches longer than the actual tail and wrap it with a fine layer of cotton (Figure 7.11). It helps to lick the wire first to get the cotton to stick. Spin the wire between your fingers as you add fine wisps of cotton until the cotton-wrapped wire is similar in shape to the tail vertebrae on the carcass; do not make it too thick. Gently insert the wire into the tail until it reaches the tip (Figure 7.11). Position the other end of the wire within the body cavity just below the mid-ventral incision. Stitch the incision closed with a needle and thread so that the incision nearly disappears.

For large mammals, it may be necessary to also create artificial legs using cotton wrapped wires. If this is the case, carefully position the base of the wires along the ventral side of the cotton "body" as the skin is reversed over it. Tie a properly labeled field tag on the right hind foot of the skin at the ankle.

The study skin must be pinned to a flat piece of cardboard or Styrofoam for drying. The front and hind feet are pinned parallel to the body (Figure 7.12). The tail is positioned so that it extends straight out from the middle of the body. Two pins angled over the tail (not through the tail) hold it in place at several points along its length. Check the overall appearance of the study skin and make adjustments to shape before gently brushing the fur with a toothbrush (to remove sawdust and dirt).

Pinning bats is more complicated because of the wings. Figure 7.12 shows the pin locations for a typical bat study skin. Pinned specimens must dry thoroughly before shipping or transporting (this can be difficult in humid rain forests).

The carcass can be prepared as a skull and partial skeleton following the procedures described in the next section.

Exercise 3: Preparing Skulls

As soon as the study skin is prepared, separate the skull from the carcass between the skull and the first vertebra (atlas). Fragile skulls from small mammals (shrews, mice, small bats) are dried without any cleaning. For larger specimens, it is necessary to remove the brain, eyes, tongue, and heavy jaw muscles. The brain is removed by forming a small U-shaped hook on the end of a wire and inserting this into the cranium through the foramen magnum. The brain tissue is broken up by gentle side-to-side movement of the wire inside the cranium. A syringe of water is used to flush out the remaining brain tissue. Remove the eyes and tongue with forceps and cut away large jaw muscles with scissors (be careful not to damage the zygomatic arch). Write the field number in pencil on a heavy paper or cardboard tag and tie it to the skull. In the field, skulls are dried in cloth or gauze bags placed in a warm ventilated area. Be sure drying skulls and skeletons are placed out of reach of animals.

Exercise 4: Preparing Skeletons

The procedures for preparing skeletons is similar to that for skulls. The skeleton is cleaned of organs, and larger muscles are removed from the bones. The amount of defleshing depends on the size of the mammal. Shrews, mice, and small bats do not usually require defleshing, and the skulls (if it is a full skeleton and skull) can be left attached. For medium-sized mammals defleshing is necessary and the skull must be separate from the skeleton and prepared as described above. After defleshing, field numbered tags are attached and the skeletons are placed in gauze bags to protect them from egg-laying flies. Do not place skeletons or skulls in plastic bags, as the lack of ventilation will cause them to decompose instead of drying. Do not spray the skeletons with insecticides as they will inhibit the activity of dermestid beetles that are used in many museums to clean skeletons.

Fresh tissues from many organs (heart, kidney, liver, ovary, testes, and muscle) can be preserved in the field by flash freezing. Small 5- to 10-mm cubes of the desired organ is removed from the animal and placed into a plastic cryotube. Small tags with field numbers and tissue type (heart, muscle) are placed inside the tube before it is sealed, and the field number is also written in permanent ink on the outside of the tube (along with species name and sex). Ideally the tissue in the cryotubes is flash frozen in liquid nitrogen or wrapped in aluminum foil and stored in dry ice. Once samples are brought to the laboratory, they can be stored in a freezer at −70°C for six months.

8   Field Collecting and Preserving Mammalian Parasites

Time Required
- Collection of ectoparasites and endoparasites can be accomplished in one 3- to 4-hour lab period or spread across two shorter periods.

Learning Objectives
- Understand the different methods for collecting parasites from mammal specimens.
- Learn how to identify types of ectoparasites.
- Learn how to prepare a blood smear.
- Learn how to record parasite data along with specimen data in the field.

Equipment Required
- Field notebook with permanent ink pen or pencil
- Precut paper labels on cardstock
- Dissection scissors (iris scissors and larger dissection scissors)
- Scalpel and blades
- Fine forceps and large blunt forceps
- Disposable plastic pipettes
- Microscope slides and coverslips
- Coplin jar
- Filter paper for DNA samples
- Methanol, 100%
- Ethanol, 95%
- Sterile cotton swabs and toothbrush
- Vials and tubes
- Large plastic bags
- Whirl-pak plastic bags
- Chloroform or ether
- Comb with fine metal tines (lice comb)
- Dissecting microscope and battery-operated light source
- Liquid nitrogen tank and cryovials
- Prepackaged potassium dichromate

Background

Mammals play host to a variety of ectoparasites and endoparasites. Several mammalian hosts and their parasites co-evolved (Hafner and Nadler, 1988; Presley et al., 2015). Thus, parasites influence host longevity, and in doing so also influence parasite density and diversity. In some cases, parasite load causes increased mortality of hosts (Hersh et al., 2014; Moore and Wilson, 2002; Morand and Harvey, 2000). In other cases, host mortality is largely unaffected by parasitic infestation. Mammalian parasites may also reduce hosts' reproductive potential (Hunter and Webster, 1974), increase the risk of disease outbreaks, and move between mammalian species (Williams and Barker, 2001).

In the past, some mammalogists considered the collection of parasites as ancillary to their own studies. However, research on host-parasite ecology and phylogeny are leading to fresh insights on how mammals evolved and how diseases spread within and among mammalian populations (Knowles et al., 2013).

The goal of the exercises that follow is to familiarize students with the basic methods for collecting and preserving mammalian parasites. Mammalogists rarely have the expertise to identify parasites to species. Rather they submit their parasite collections to experts in each taxonomic group for identification. Therefore, only a very basic identification is discussed here.

Exercise 1: Making a Blood Smear

Mammals host a number of hemoparasitic species including trypanosomes and plasmodium. Parasitologists and epidemiologists can use blood smears collected in the field to study transmission within and between species.

1. On a recently euthanized mammal, collect several drops of blood with a 30-70 microliter (µl) heparinized capillary tube. Blot the blood from the tube on filter paper. Clearly label the filter paper with the mammal's field number, collector, species name, and sex. Store the filter papers in a cool dry place. The filter paper blood sample will be used for DNA analysis later.

2. Using the remaining blood, prepare a single smear on a microscope slide (Figure 8.1).

3. Use microscope slides with a frosted end, so that identifying information can be written there in pencil. Place a drop of blood approximately 4 mm in diameter near the end of the slide.

4. To spread the drop, use the edge of another slide ("spreader" slide). Place the spreader slide at a 45° angle with the bottom edge just touching the drop of blood. The spreader uses capillary action to drag the droplet. Push the spreader across the slide dragging the blood across the surface of the slide with it. Use a smooth quick action to make a nice, even smear.

5. Air-dry the smears completely, and then dip them into 100% methanol in a Coplin jar for one to two minutes.

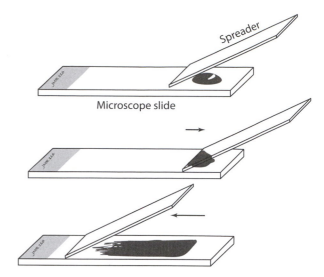

Figure 8.1 How to perform a blood smear. The spreader slide is used to quickly move the blood across the slide surface.

6. After methanol fixation, remove the slides and blot the end on filter paper to drain excess alcohol. Store the slides in a plastic slide box for complete drying and transporting.

7. Send smear specimens to a parasite specialist as soon as possible after returning from the field so that they can be stained and analyzed.

Exercise 2: Collecting Ectoparasites

1. Place a recently dead small mammal specimen in a new plastic bag (do not reuse bags).

2. Record the host animal's field number on labels that will stay with the parasites.

3. Place the bagged specimen in a large Tupperware container containing chloroform- or ether-soaked cotton in a well-ventilated area. Close the plastic lid but do not seal the plastic bag inside. It is used to catch any parasites that fall off the host.

4. Allow sufficient time for any ectoparasites to die, and without removing the mammal from the plastic bag, carefully brush the fur with a toothbrush so that any parasites fall into the bottom of the bag.

5. Carefully examine the pinna and body surface for any additional parasites. Then remove the mammal from the plastic bag and examine the fur under a dissecting

microscope to remove any additional parasites. When finished, save the mammal for endoparasite collection or for depositing into a mammal collection.

6. Place a parasite label tag inside the bag along with any additional parasites found during microscope inspection.

7. Add 5-8 milliliters (ml) of 70% ethanol to the plastic bag, and shake all parasites and fluid to one corner. Hold the fluid corner over a small vial and cut the corner with scissors so that it drains into the vial. Be sure the vial is large enough to contain all the fluid.

8. Once the parasites have been fixed in ethanol, spread them out in a petri dish for sorting and identification under a dissecting microscope. Be sure to keep the specimen tag with the samples at all times.

Exercise 3: Collecting Endoparasites

Endoparasites occur in many species of mammal. They reside in a variety of body organs, most notably in the gastrointestinal (GI) tract, liver, and lungs. They may include various species of tapeworms, flukes, nematodes, coccidia, trypanosomes, and protozoans.

1. Prepare a blood smear on a standard microscope slide as described above. The smear can be stained and examined for trypanosomes and blood protozoans when time permits.

2. A sample of feces is saved in a vial half-filled with a 2% potassium dichromate solution. Feces can often be removed from the anus with forceps if necessary. Care must be taken to ensure that the sample is not cross-contaminated with those of other individuals (it is not a good idea to remove fecal pellets from traps). Insert a specimen tag into each vial.

3. To collect endoparasites from the digestive system (GI tract), open the abdominal cavity and sever the esophagus just above the stomach and the colon just in front of the rectum. Do not cut into the organs.

4. Remove the intact GI tract, and place it in a clean petri dish or enamel pan (with a specimen tag) containing water or saline.

5. Examine the body cavity, liver, kidneys, and lungs for helminth cysts or worms. Check the heart, aorta, pleural cavity, mesenteries, or subcutaneous tissues for filarial worms and other parasites.

6. After searching the body cavity, return to the GI tract. Open each organ separately, and examine for parasites under a dissecting microscope.

7. It is important to relax the parasite specimens before fixation. To relax a tapeworm, trematode, or platyhelminth, place the worm in water. This creates an osmotic imbalance, which causes death and the eversion of the rostellum or proboscis, which is often necessary for identification. The relaxation process can take a few minutes to over one hour, depending on the size and species of the worm.

8. After relaxation, the parasites are preserved in 10% buffered formalin solution (see Chapter 7) and placed in vials with both the field number and the location of parasite on the tag (L for liver, S for stomach, SI for small intestine).

Exercise 4: Preliminary Ectoparasite Identification

Researchers usually send their parasite samples to experts to get definitive species-level identifications. However, it is possible to identify many parasites to major group. Sort the ectoparasites into the major groups listed below.

Ticks

Ticks are arachnids and along with mites form the subclass Acari. Ticks are vectors of several important diseases, including Lyme disease, Rocky Mountain spotted fever, and typhus.

Most ticks belong to one of two major groups, the hard ticks (Ixodidae) and the soft ticks (Argasidae). Adult hard ticks have pear-shaped bodies, eight legs, and beak-like mouthparts in front (Figure 8.2). Male hard ticks have a hard dorsal shield (scutum) covering most of the body, while females have a shield covering only the anterior third or so of the back. Many hard ticks have a multi-host life cycle (Figure 8.3), where the tick takes a blood meal at each developmental stage. Typically, the female tick, and not the male, takes the blood meal.

Soft ticks lack a dorsal shield, and their mouthparts are hidden on the underside of the head when viewed from above (Figure 8.4). Adults are oval and dorsoventrally flattened. Their dorsum is typically very rough or granulated. Soft ticks are intermittent feeders and do not remain attached to the host for long periods. Several soft ticks in the genera *Argas* and *Carios* feed on bats and birds.

There are more than 700 species of ticks in 14 genera. In the United States, ticks typically belong to four genera: *Amblyomma*, *Dermacentor*, *Ixodes*, and *Rhipicephalus*. *Amblyomma americanum* is commonly known as the Lone Star tick. Adults are identified by the light patch in the middle of the back. The genus *Dermacentor* includes two common species: *D. variabilis* (American dog tick) and *D. andersoni* (Rocky Mountain wood tick). These ticks have variable patterns of cream color on their dorsal shields. Deer ticks (also called black-legged ticks) belong to the genus *Ixodes*. They are identified by their reddish body and dark brown dorsal shield (without lighter markings). Finally, *Rhipicephalus sanguineus* or the brown dog tick has a more uniform brown dorsum.

Mites

The second member of the Acari subclass are mites. More than 48,000 species have been described so far, and many more probably remain to be identified. Because mites are so small, they are typically mounted on a

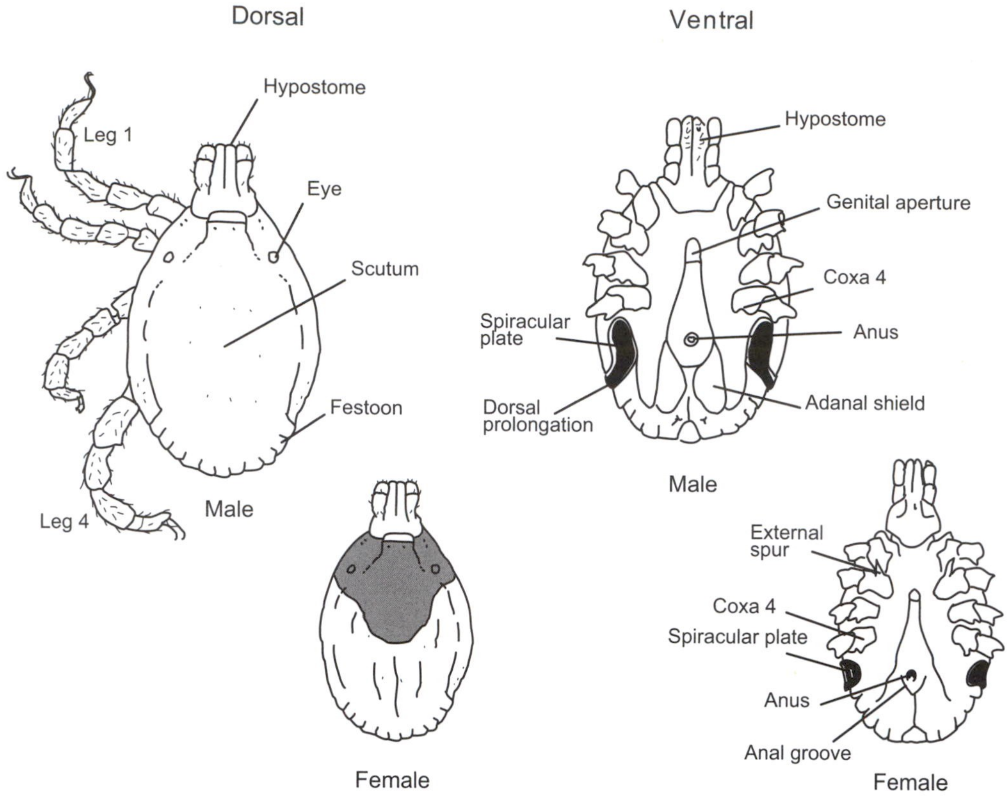

Figure 8.2
Anatomy of a hard tick. Dorsal views are on the left, and females are shown in smaller drawings along the bottom.

Dorsal

Ventral

Hypostome

Leg 1

Eye

Scutum

Leg 4

Male

Festoon

Female

Hypostome

Genital aperture

Coxa 4

Spiracular plate

Anus

Dorsal prolongation

Adanal shield

Male

External spur

Coxa 4

Spiracular plate

Anus

Anal groove

Female

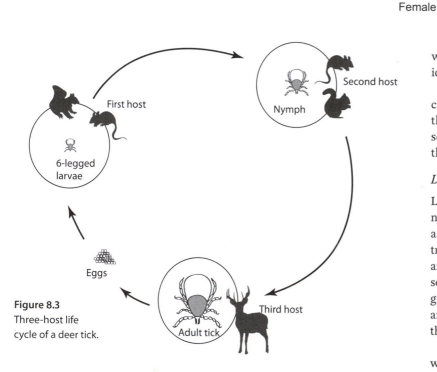

First host

6-legged larvae

Eggs

Second host

Nymph

Third host

Adult tick

Figure 8.3
Three-host life cycle of a deer tick.

which reside in hair follicles and sebaceous glands, are identified by their slender, elongate bodies.

Chiggers are mites belonging to the family Trombiculidae. These mites do not burrow into the skin. Instead they insert their mouthparts into the skin and inject a secretion that dissolves the surrounding tissue so that they may feed on it.

Lice

Lice belong to the order Phthiraptera, which contains nearly 5,000 species of apterous (wingless), obligate parasites. Mammalian lice are soft bodied and dorsoventrally flattened (Figure 8.6A). They possess small eyes and antennae, have sucking or chewing mouthparts, and some have hook-like claws at the end of their legs for grasping hair shafts. Sucking lice pierce the host's skin and suck in blood and other fluids. Chewing lice live in the host's fur, feeding on dead skin and cellular debris.

Lice are hemimetabolous. The eggs hatch into nymphs, which resemble smaller versions of the adult. The nymphs molt three times over the course of four weeks before becoming fully adult (Figure 8.6B).

Fleas

The order Siphonaptera (fleas) are a group of approximately 2,500 species of obligate parasitic insects commonly known as fleas. Adult fleas are 2 to 5 mm in length and have laterally compressed bodies, allowing them to move easily among the host's hair. They lack wings and instead use their large hind legs to jump from one

microscope slide for identification. Mites include several important species of mammalian ectoparasites, including the scabies mite (*Sarcoptes scabiei*), the mouse mite (*Allodermanyssus sanguineus*), and chiggers (trombiculid mites).

Adult mites are 0.2 to 0.4 mm long, with rudimentary legs (Figure 8.5). The body is covered with spines and long hair-like structures (setae). Follicle mites (*Demodex*), patch of fur to another or from one host to another (Figure 8.7). Fleas have sucking mouthparts that are directed downward.

Fleas undergo complete metamorphosis (holo-metabolis)—that is, larvae and adults are morphologically different. Only adults are parasitic. Eggs are deposited in the host's nest or burrow, where they hatch into maggot-like larvae. The larvae feed on feces and other organic material before they spin a cocoon and pupate into adults.

Bot flies

Bot flies (also known as warble flies) belong to the order Diptera (family Oestridae) along with many other species of fly. Bot flies deposit eggs on the host (or another vector of that host), and the larvae grow inside the host

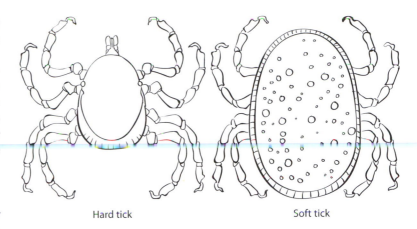

Hard tick   Soft tick

Figure 8.4 Morphology of a hard (*left*) and soft (*right*) tick shown in dorsal view. Notice that the mouthparts of the soft tick are completely hidden from view.

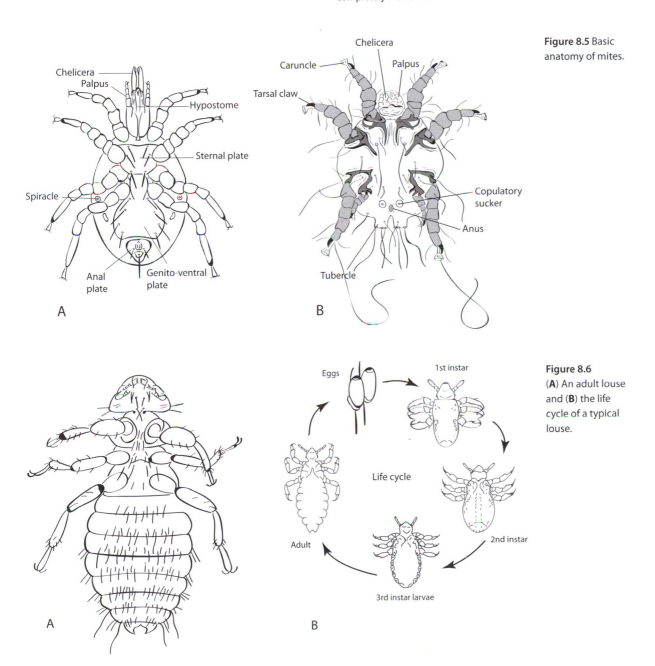

Figure 8.5 Basic anatomy of mites.

Figure 8.6 (**A**) An adult louse and (**B**) the life cycle of a typical louse.

Figure 8.7 Scanning electron micrograph of a flea. *Centers for Disease Control and Prevention (CDC) Public Health Image Library/Janice Haney Carr*

animal's flesh. Developing larvae maintain a breathing tube, which opens at the skin's surface. In mice, bot fly larvae typically inhabit the abdominal or inguinal tissue. This has led to the notion that bot fly maggots emasculate their hosts. However, Timm and Lee (1981) showed that larvae-infested mice are not emasculated and do not suffer reduced fertility after infestation.

Bat flies

Another group of flies in the order Diptera infests bats. Bat flies belong to the families Nycteribiidae and

Figure 8.8 A parasitic bat fly (family Nycteribiidae) collected on a *Plecotus auritus* bat in Switzerland. *Courtesy of Gilles San Martin from Namur, Belgium*

Streblidae. Bat flies have a spider-like appearance with a somewhat flattened body and laterally extended legs (Figure 8.8). Unlike spiders, bat flies have three pairs of legs. Most species lack eyes and wings. They feed on the host's blood.

Most bat flies reproduce in an unusual manner. Eggs are fertilized internally, and females retain these eggs within the body. All larval stages develop inside the mother's body nourished by secretions from intrauterine glands ("milk" glands). Following two larval molts, the third instar larva is deposited on the substrate near the host. This third instar larva immediately pupates, and within four weeks a wingless adult emerges. Newly emerged adults walk around until they encounter a host (Dick and Patterson, 2006).

Bat flies are highly host-specific (Dick and Gettinger, 2005) and rarely leave their host's body. Because of this, bat flies have been used to elucidate the evolution of several groups of bats (Dittmar et al., 2006; Patterson et al., 2009).

Exercise 5: Ectoparasite Population Ecology

Identifying ectoparasites is the first step in understanding the ecology of host-parasite interactions. After collecting ectoparasites from a population of small mammals, record the information below and write a paper using some or all of the analyses below.

1. For each major group of ectoparasites encountered, describe the location of the parasites on the host. Do they cluster in one region of the host's body? Are the same types of parasites found in different places on different hosts?

2. Calculate the prevalence, mean abundance, and mean intensity of each type of parasite as described below:

$$Prevalence = \frac{(number\ of\ animals\ with\ parasites)}{(number\ of\ animals\ examined)} \times 100$$

$$Mean\ abundance = \frac{\left(\begin{array}{c}total\ number\ of\ parasites \\ of\ a\ particular\ type\end{array}\right)}{(number\ of\ animals\ examined)}$$

$$Mean\ intensity = \frac{\left(\begin{array}{c}total\ number\ of\ individual \\ parasites\ of\ a\ particular\ species\end{array}\right)}{\left(\begin{array}{c}number\ of\ animals\ examined \\ that\ were\ infested\ with\ that \\ parasite\ species\end{array}\right)}$$

3. If available, plot ectoparasite burdens over time (periods of one or more years).

4. One of the problems with reporting mean abundance and mean intensity is that most hosts have few if any parasites and a few hosts are heavily infested. This leads to a right-skewed distribution, which invalidates the assumptions of many traditional statistical methods. To avoid this problem, use the QPWeb Quantitative Parasitology website to perform statistical analyses (Reiczigel et al., 2013; Rozsa et al., 2000; http://www2.univet.hu /qpweb/).

5. Explore a real data set of rodent ectoparasites from the Sevilleta National Wildlife Refuge in New Mexico. These data are available at http://sev.lternet.edu/data /sev-13, and in the data files that accompany this text. Over a nine-year period, researchers collected ectoparasites from 3,235 rodents in 28 species from three habitat communities. The metadata on this site provide a key to abbreviations for each category. Be sure to look over the metadata before starting. The data are available to download as a .txt file that can be imported into Excel for analysis. Choose one species of small mammal from the study. Import the data into Excel and answer the following questions:

- Do parasite intensities change over time (year-to-year or within a year) for your species of mammal?
- Do males or females of your species have higher parasite loads?
- Which parasite group is most abundant on your mammal species?
- Which tissue type harbors the most parasites in your species?
- Are there seasonal differences in parasite load?
- Are there differences in parasite load between sites for your species?

6. Write up your findings in *Journal of Mammalogy* format as if you planned to submit it for publication. The guidelines for authors can be found at https://academic .oup.com/jmammal/pages/general_instructions.

9 Mark-Recapture Studies

Time Required
- The simulation may be completed in a typical 3- to 4-hour lab period.

Learning Objectives
- Understand the principles of mark-recapture studies.
- Understand the difference between open and closed population methods.
- Learn how to use the Lincoln-Petersen method.
- Learn how to use the Schnabel model.
- Learn how to use the Jolly-Seber model.

Equipment Required
- Plastic containers of white beans
- Plastic containers of dark beans
- Data sheets
- Calculators

Background

How many moose are in the Adirondack Mountains of New York State? How many elephants are in the Serengeti ecosystem? How many voles are there in a hectare of boreal forest in northern Canada? The answers to these questions require knowledge of the abundance of the mammals in question. Estimating abundance is a very common procedure for ecologists and conservation biologists. After all, population size can influence many other population parameters, including the genetic composition, rates of immigration, emigration, birth, and long-term survival of the population in a given region.

Population size is determined in two ways. The first is an actual census or head count of all individuals within a specified region. Often this is impractical or impossible. A second approach is to estimate population size through subsampling. Census techniques include aerial surveys, where the pilot flies in a grid pattern over the study area and an observer counts all individuals sighted. Aerial surveys work well for large mammals in relatively open habitats; elephants, giraffes, wild horses, and other large mammals are often counted using aerial census techniques. Obviously, direct counts of many small mammals, nocturnal or cryptic species, fossorial (burrowing) mammals, or species inhabiting densely vegetated areas are nearly impossible. In such cases, mammalogists rely on estimates of population size based upon a subsample of the entire population.

When direct census methods cannot be used, abundance is often estimated using mark-recapture techniques (Lancia et al., 1994; Sutherland, 1996). Large mammals, such as elk or deer, can be marked with ear tags, and small mammals can be marked by using tiny ear tags, clipping fur, or dying small patches of fur (see Chapter 6). In its simplest form, there are two sessions. The first session involves capturing a subset of the population, marking them, and releasing them. Marked animals then move freely and intermix with unmarked members of the population. A second session involves capturing a random sample of individuals from the population and determining how many of them were marked in the first capture session. More complex techniques involve multiple recapture sessions and uniquely marking each individual. Mark-recapture techniques are based on the idea that the proportion of marked individuals in the second sample should be approximately equal to the proportion of marked animals in the total population. Thus, knowing the number of marked and unmarked animals captured in the second session, and the number originally marked in the first session, allows an estimate of the total population size.

This chapter presents three different mark-recapture models: the Lincoln-Petersen model, the Schnabel model, and the Jolly-Seber model.

Lincoln-Petersen Method

The Lincoln-Petersen method is the simplest because it requires only two sessions: a single marking session and a single recapture session (Seber, 1982). The data in the model include the number of individuals marked in the first sample (M); the total number of individuals captured in the second sample (C); and the number of individuals in the second sample that have marks from the first session (R). These data are used to estimate total population size, N, as

$$\frac{N}{M} \approx \frac{C}{R}$$

If the task is to estimate the population of voles in a given area, then this equation says that the ratio of the total number of voles in the population to the total number of marked voles is approximately equal to the ratio of the number of voles in the second sample to the number of marked (recaptured) voles in the second sample. Solving for N, an estimate of the total population size yields

$$N = \frac{CM}{R}$$

This formula is the Lincoln-Petersen index of population size. More commonly this equation is modified as

$$N = \frac{(M+1)(C+1)}{R+1} - 1$$

The Lincoln-Petersen estimate assumes that

- the population is closed (no immigration or emigration and no births or deaths during the study),
- the population does not change in size between the mark and recapture sessions,
- the second sample is a random sample,
- marking does not affect the recapture of individuals, and
- marks are not lost, gained, or overlooked during the study.

Let us estimate a hypothetical population of voles using sample data. Suppose you set up a large livetrap study. The following morning you return to the trapping grid to check your traps, and you mark all of the individuals captured by clipping the fur on their flank. You then release each captured animal at its capture site and rebait and reset the traps. The next day (session 2) you revisit the grid in the morning and record the number and type (marked or unmarked) of each capture. Your data are as follows: $M = 99$, $C = 149$, and $R = 24$.

Entering the data into the Lincoln-Petersen equation gives the following:

$$N = \frac{(99+1)(149+1)}{24+1} - 1 = \frac{100(150)}{25} - 1 = 599$$

According to this, there are an estimated 599 voles at the study site. Of course, we would like to know whether

this number is accurate or not. One way to determine accuracy is to calculate the approximate confidence limits of the estimate.

First, calculate p:

$$p = R/C = 24/149 = 0.16$$

Next, calculate the upper and lower confidence limits:

$$W_1, W_2 = p \pm \left[1.96 \sqrt{\frac{p(1-p)\left(1-\frac{R}{M}\right)}{(C-1)+\frac{1}{2C}}} \right]$$

$$W_1, W_2 = 0.16 \pm \left[1.96 \sqrt{\frac{0.16(1-0.16)\left(1-\frac{24}{99}\right)}{(149-1)+\frac{1}{2(149)}}} \right]$$

$$W_1, W_2 = 0.16 \pm \left[1.96 \sqrt{\frac{0.16(1-0.16)(1-0.24)}{(148)+\frac{1}{298}}} \right]$$

$$W_1, W_2 = 0.16 \pm \left[1.96 \sqrt{\frac{0.16(0.84)(0.76)}{(148)+0.00336}} \right]$$

$$W_1, W_2 = 0.16 \pm \left[1.96 \sqrt{\frac{0.102}{148}} \right]$$

$$W_1, W_2 = 0.16 \pm 0.0515$$

Divide M by W_1 and by W_2 to give the approximate 95% confidence limits for your calculated value of N (599 voles).

$$W_1 = \frac{99}{0.21} = 471 \text{ voles}$$

$$W_2 = \frac{99}{0.11} = 900 \text{ voles}$$

In this example the Lincoln-Petersen estimate gives us a vole population of 599 voles, and we can be reasonably sure that the actual population is somewhere between 471 voles and 900 voles.

Schnabel Model

The Schnabel model is similar, in theory, to the Lincoln-Petersen method but involves more than one mark-and-recapture session. This method relies on the same assumptions as before. In the Schnabel method, every

Table 9.1 Five days of mark-recapture data used to illustrate the Schnabel method

i	C_i	R_i	U_i	M_i
1	10	0	10	0
2	15	5	10	10
3	10	2	8	20
4	5	0	5	28
5	5	4	1	33

Table 9.2 Summary values calculated from the values in Table 9.1

$C_i M_i$	R_i
0	0
150	5
200	2
140	0
165	4
A = 655	B = 11

time you sample the population, you mark any unmarked captures and release them all back into the habitat. Thus, the proportion of marked animals in the population increases with each sampling session. When this proportion is equal to 1.0, every individual has been marked and the population size is therefore equal to the total number marked. Obviously, you do not need to continue sampling until every individual is marked, but you need a way of modeling the change in that proportion.

Let us apply the Schnabel method to our hypothetical vole population. Assume you have the following:

S = number of sample sessions
C_i = number of animals in the ith sample
R_i = number of marked animals in the ith sample (recaptures)
U_i = number of animals marked for the first time and released in the ith sample
M_i = number of animals marked prior to the ith sample – or $\sum U_i$

Suppose you sample the trapping grid on five separate days ($S = 5$), and you have a set of mark and recapture data as shown in Table 9.1.

The Schnabel method is essentially a series of Lincoln-Petersen samples. The Schnabel formula is:

$$N = \frac{\sum(C_i M_i)}{\sum R_i}$$

Using the data in the Table 9.1, we can calculate the numerator and denominator as shown in Table 9.2.

Using the equation above for N, the estimate of the total population size is

$$N = \frac{655}{11} = 59.5$$

The 95% confidence limits would be calculated using a Poisson distribution table (such tables can be found in most statistics texts or online at http://statpages.info/confint.html). In this case we have 11 total recaptures so we check the Poisson table for x = 11 and see upper and lower 95% confidence intervals of 19.68 and 5.49, respectively. Using the Schnabel equation again, we substitute the confidence values for the denominator in the equation.

$$N = \frac{\sum(C_i M_i)}{5.49} = \frac{655}{5.49} = 119 \text{ and } N = \frac{\sum(C_i M_i)}{19.68}$$
$$= \frac{655}{19.68} = 33$$

Thus, using the Schnabel method our new animal population is estimated to be 60 animals, or between 33 and 119 animals.

Jolly-Seber Model

The Jolly-Seber model does not require a closed population. Animals are assumed to be able to move in and out of the population. This method requires knowledge of survival between samples and is mathematically complex. The Jolly-Seber model includes information on when a marked individual was last captured. This requires marks or tags that are unique to the animal and records of the dates that each animal was captured. Beyond the advantage of assuming an open population, the time intervals between sampling periods do not need to be constant. This means that a population may be sampled over many months or several years. A minimum of three sampling periods are required (more are better).

Like the models described previously, the Jolly-Seber model makes several important assumptions (Krebs, 1999):

- Every individual has the same probability of being captured in the ith sample (whether it is marked or unmarked).
- Every marked individual has the same probability of surviving from one sampling period to the next.
- Individuals do not lose their marks, and marks are not overlooked at capture.
- Sampling time is negligible in relation to the intervals between samples.

The variables required for the Jolly-Seber model are as follows:

- m_t = Number of marked animals caught in sample t
- u_t = Number of unmarked animals caught in sample t
- n_t = Total number of animals caught in sample t ($n_t = m_t + u_t$)
- s_t = Total number of animals released after sample t
- m_{rt} = Number of marked animals caught in sample t, last caught in sample r
- R_t = Number of the s_t individuals released after sample t and caught again in some later sample

- Z_t = Number of individuals marked before sample t, not caught in sample t, but caught in some sample after sample t

The three equations are as follows:

$$\bar{\alpha} = \frac{m_t - 1}{n_t - 1}$$

where alpha is an estimate of the proportion marked at time t.

$$M_t = \frac{(s_t + 1)Z_t}{R_t + 1} + m_t$$

where M_t is an estimate of the marked population just before sample time t.

$$N_t = \frac{M_t}{\bar{\alpha}_t}$$

where N_t is an estimate of the population size at time t.

In Table 9.3, the diagonal portion of data in the upper right of the table represents the actual number of marked voles collected over seven sampling sessions, each spaced two weeks apart. During the first sampling session 22 voles were caught and marked, and 21 voles were released (e.g., one vole may have died in the trap). At sampling session 2, 41 voles were caught, and of those 15 were marked in the first sampling session (15 in upper box for column 2). The remaining 26 unmarked voles were marked and released at the end of sampling session 2. At sampling session 3, 48 voles were captured. Of those 16 had marks (15 were originally marked in the last sampling session 2 and 1 was marked last in sampling session 1. This illustrates an important point about the Jolly-Seber model: marked voles may not be captured in every session; some marked voles may evade capture for one or more sessions.

In Table 9.4, take a look at the sampling session 5 (column 5): 89 voles were captured, of which 64 were marked previously in either session 4 (61 of the marked voles), in session 3 (2 of the marked voles), or in session 2 (1 vole). No voles from session 1 were recaptured in session 5. Of the 89 voles captured in session 5, only 88 were released (s_t). Now take a look at the horizontal row for session 4 (beginning with 61). This row illustrates the number of voles released at session 4 that were captured again in a later session. Thus, of the 61 voles marked and released at session 5, four of those voles were captured again in session 6, and one vole in session 7. Finally, the values in bold font represent the voles captured prior to session 4, but not in session 4, that were captured in a later session (e.g., for session 4 this would be two voles). Using the values in the four rows at the bottom, we can calculate the following summary data (Table 9.5):

Table 9.3 A set of hypothetical mark-recapture data for a series of capture sessions of meadow voles (*Microtus pennsylvanicus*)

Time of last capture	Time of capture						
	1	2	3	4	5	6	7
1		15	1	1	0	0	0
2			15	2	1	0	1
3				37	2	2	0
4					61	4	1
5						77	3
6							77
Total marked (m_t)	0	15	16	40	64	81	82
Total unmarked (u_t)	22	26	32	45	25	22	26
Total captured ($n_t = m_t + u_t$)	22	41	48	85	89	103	108
Total released (s_t)	21	41	46	82	88	99	106

Table 9.4 Highlights of the data presented in Table 9.3

Time of last capture	Time of capture						
	1	2	3	4	5	6	7
1		15	1	1	0	0	0
2			15	2	1	0	1
3				37	2	2	0
4					61	4	1
5						77	3
6							77
Total marked (m_t)	0	15	16	40	**64**	81	**82**
Total unmarked (u_t)	22	26	32	45	25	22	26
Total captured ($n_t = m_t + u_t$)	22	41	48	85	**89**	103	**108**
Total released (s_t)	21	41	46	82	**88**	99	106

The proportion marked in session 2 is 0.381, which we calculate as

$$\frac{m_2 + 1}{n_2 + 1}$$

or 38% of the voles in the population are marked at session 2.

The size of the marked population at session 6 in Table 9.4 is 87.7, which we calculate as

$$M_t = \frac{(n_t + 1)Z_t}{R_t + 1} + m_t = \frac{(103 + 1)5}{77 + 1} + 81 = 87.7$$

Table 9.5 Summary data, including population size estimates, for the *Microtus* population data from Table 9.3

Sample	Proportion marked (a_t)	Size of marked population (M_t)	Population estimate (N_t)
1	0.000	0.0	NA
2	0.381	19.2	50.4
3	0.347	21.8	62.9
4	0.477	47.7	100.1
5	0.722	73.1	101.2
6	0.788	87.7	111.2
7	0.761	82.0	107.7

Note: NA = not available.

where Z_t is the number of individuals marked before sample t that were not caught in sample t but that were caught again after sample t. In our case there are 5 animals in the top five rows of column 7, so $Z_6 = 5$ animals. The number of animals released in session t and caught again later is denoted by R_t. In our case, $R_6 = 77$ animals. The population size estimate at session 6 is 111.2 voles, which we calculate as

$$N_t = \frac{M_t}{\bar{\alpha}_t} = \frac{87.7}{0.788} = 111.2$$

Exercise 1: Single Mark-Recapture (Lincoln-Petersen Method)

You can simulate a mark-recapture study in the laboratory using model organisms. Here you will use beans as model organisms. The objective of this exercise is to estimate the population size (using the Lincoln-Petersen and Schnabel models, along with the 95% confidence limits). These measures are to be calculated for each trial. For the 95% confidence interval, there is a 95% certainty that the actual value for population size falls between the lower confidence limit and the upper confidence limit. Thus, more accurate predictions will have narrower confidence intervals.

To simplify the procedure for instructional purposes, you will use white beans as your model organism. Each group will have a container of white beans and a container of black beans to work with. Follow the procedure here to collect your data:

1. Put three to six handfuls of white beans into a cloth sack (this represents the population of animals from which you will be sampling). Do not count them yet. Now make a guess about how many beans you just placed in the sack, and record this guess in the worksheet provided in the appendix to this chapter.

2. Take a small handful of white beans back out of your sack. This represents your first sampling session of a group (M). Count these beans and record the number as your value for M in your data table. Do not return these beans to your sack.

3. You will now mark the beans (organisms) you just captured. To mark these beans, merely replace them with dark-colored beans. (For example, if you "captured"

25 white beans, set them aside and count out 25 dark beans to serve as your marked beans.) This simulates marking them with a colored tag of some kind.

4. Now release the marked individuals back into the "population" by placing the dark beans you counted out in step 3 into the sack. The white beans that you replaced should be returned to the original white bean container (not put back into the sack). Why?

5. Shake the sack to disperse the beans in the population. Without looking, grab a small handful of beans from the sack. This represents your second sampling session of a group of organisms (C). Count the total number of beans you grabbed in this handful (regardless of color), and record your answer as the value for C in your data table.

6. Examine the same handful of beans you gathered in step 5. Count the number of those beans that were "marked" (dark). Record this number as your value for R in the table. When you are finished counting, return this entire sample to your sack (both the white and dark beans).

7. Use the Lincoln-Petersen equation to calculate your population estimate (N). Record your answer as the value for N in your data table. Also calculate the standard error and 95% confidence intervals for your estimates.

8. Now count the actual total number of beans (both white and dark) in your sack. Record your count in the worksheet.

9. Separate the light-colored beans from the dark-colored beans, and return them to their original containers.

Exercise 2: The Schnabel Method

1. Put four to six handfuls of white beans into a sack. Do not count them. Now make a guess about how many light-colored beans you just placed in the sack, and record this guess in the Schnabel data worksheet in the appendix to this chapter.

2. Now take a small handful of white beans back out of your sack. This represents your first capture of a group of organisms (C_1). Count these beans and record the number as your value for C_1 for sampling session 1 in your data table. This also represents the number of unmarked animals in session 1, so place this value in U_1 in the worksheet. Do not return these beans to your sack. Because this is the first session, there are no recaptures (R_1) or marked individuals (M_1) yet, so place a 0 in each of those columns for session 1.

3. You will now mark the organisms (beans) you just captured. To mark these beans, merely replace them with dark-colored beans. The number of beans you marked now becomes the number of marked individuals in the population for your next sample. Record this number of marked beans as the value for M_2 for trapping session 2 in your data table (it will be the same number as for C_1).

4. Release the marked individuals back into the "population" by placing the dark beans you counted out in step 3 into the sack. This represents the number of marked animals prior to the next sample or M_2 in the worksheet. The white beans that you replaced with dark beans are set aside and never returned to the sack.

5. Shake the sack.

6. Without looking, grab a small handful of beans from the sack. This represents your second sampling of a group of organisms. Count the total number of beans you grabbed in this handful (regardless of color), and record your answer as your value for C_2 in your table. Examine this handful, determine the number of marked (dark) beans, and record this number as your value for R_2. Subtract R_2 from C_2 to get the number of unmarked animals in session 2, and place this value in U_2.

7. Still working with the same handful of beans collected in step 6, mark the unmarked (white) beans in the sample by replacing them with dark beans (you are marking previously unmarked beans). Add the number of individuals you just marked (U_2) to the number marked prior to this session (M_2), and record the resulting sum as the value for M_3 for sampling session 3. This represents the total number of marked individuals now in the population. Return all the beans from this second collection (which are now all marked, and therefore dark) to the sack.

8. Shake the sack.

9. Without looking, grab a small handful of beans from the sack. This represents your third sampling session. Count the total number of beans you grabbed in this handful (regardless of color), and record your answer as your value for C_3. Examine this handful, determine the number of marked (dark) beans, and record this number as your value for R_3. Subtract R_3 from C_3 to get the number of unmarked animals in session 3, and place this value in U_3.

10. Still working with the same handful of beans collected in step 9, mark the unmarked (white) beans in the sample by replacing them with dark beans (you are marking previously unmarked beans). Add the number of individuals you just marked to M_3, and record the resulting sum as the value for M_4 for sampling session 4. This represents the total number of marked individuals now in the population. Return all the beans from this third collection (which are now all marked, and therefore dark) to the sack.

11. Shake the sack.

12. Without looking, grab a small handful of beans from the sack. This represents your fourth sampling of a group of organisms. Count the total number of beans you grabbed in this handful (regardless of color), and record your answer as your value for C_4 in your table. Examine this handful, determine the number of marked (dark) beans, and record this number as your value for R_4. Subtract R_4 from C_4 to get the number of unmarked animals in session 4, and place this value in U_4. Repeat the procedures above until you have five sampling sessions.

13. Calculate the population estimate (N) using the Schnabel estimate, and record your answer in your data table.

14. Now count the actual total number of beans (both white and dark) in your sack. Record your count as the actual population size.

15. Separate the white beans from the dark beans, and return them to their original containers.

16. Calculate the Schnabel estimate and the 95% confidence limits for these data. Answer the following questions:

- For the simple mark-recapture in Exercise 1, how did your initial guess of the population size compare with the actual population number determined by a direct count?
- For the simple mark-recapture in Exercise 1, how did your initial guess of the population size compare with the calculated population estimate (N)?
- For the repeated mark-recapture in Exercise 2, how did your initial guess of the population size

compare with the actual population number determined by a direct count?
- For the repeated mark-recapture in Exercise 2, how did your initial guess of the population size compare with the calculated population estimate (N)?

- Did the simple mark-recapture or the repeated mark-recapture provide the most accurate estimate of population size (N)?
- How accurate would these estimates be if you took a pinch of beans in your sample instead of a small handful? Why?

Exercise 3: The Jolly-Seber Model Using Excel

The Jolly-Seber model can be calculated by hand, but it is a complicated process if the number of sessions is large. To make the calculations a bit easier, you will put a set of sample data into a spreadsheet (Microsoft Excel or OpenOffice). Your instructor will show you the basic operation of a spreadsheet, and how to enter formulas into a cell in a spreadsheet. Follow the procedures outlined here:

1. Enter the data for the capture histories for a small mark-recapture study of voles (*Microtus*) as shown in Table 9.6. Be sure to enter the data in each cell exactly as shown in Table 9.6. For example, the number 14 in session 2 is in cell D5.

2. Below the capture histories, type in the data for the first column of capture data and the number of unmarked animals in each session (u_t) as shown below in Table 9.7.

For the remainder of values in the "Total marked (m_t) row," we will enter a formula that adds the values from the session above. Click on cell E15, type in the formula =SUM(E5:E14), and hit return. Now grab the lower left-hand corner of this cell (E15) (the cursor turns to a +), and drag it horizontally to the right as far as cell H15. This will copy the SUM formula into the remaining cells in that row. It should now look like the line in Table 9.8.

Notice that the formula adds the values in the capture history table above for each capture session (column).

Table 9.6 A Jolly-Seber mark-recapture table for a study of *Microtus* over six trapping sessions

	A	B	C	D	E	F	G	H
1								
2								
3		Time of last capture		Time of capture				
4			1	2	3	4	5	6
5		1		14	8	7	4	2
6		2			5	4	2	1
7		3				3	2	4
8		4					14	7
9		5						12
10		6						

Table 9.7 Capture histories for the *Microtus* data set

	A	B	C	D	E	F	G	H
14								
15	Total marked (m_t)		0	14				
16	Total unmarked (u_t)		193	25	29	38	34	21
17	Total caught ($n_t = m_t + u_t$)		193					
18	Total released (s_t)		193					

Table 9.8 Total marked *Microtus*

Total marked (m_t)	0	14	13	14	22	26

Table 9.9 The completed Jolly-Seber data table for the *Microtus* trapping sessions

	A	B	C	D	E	F	G	H
1								
2								
3		Time of last capture		Time of capture				
4			1	2	3	4	5	6
5		1		14	8	7	4	2
6		2			5	4	2	1
7		3				3	2	4
8		4					14	7
9		5						12
10		6						
11		7						
12		8						
13		9						
14		10						
15	Total marked (m_t)		0	14	13	14	22	26
16	Total unmarked (u_t)		193	25	29	38	34	21
17	Total caught (n_t)		193	39	42	52	56	47
18	Total released (s_t)		193	39	42	52	56	47

3. In cell D17 type in the formula **=SUM(D15:D16)** and hit return. This simply adds the unmarked animals to the marked animals to get a total number of animals for that session. Again, grab the lower left-hand corner of cell D17, and drag it horizontally to the right as far as cell H17. This will copy the SUM formula into the remaining cells in that row. Now type in the same values for s_t as you calculated for n_t. This simply means that you marked all the captures and released them all with no deaths. It should now look like Table 9.9.

Now we want to calculate the proportion marked (α_t), the size of the marked population (M_t), the population estimate (N_t), and the probability of survival (ϕ_t). To do this using the spreadsheet, you will set up a second table below the first table. The new table will begin with cell A24. Type the five column headings as follows:

In cell A24 enter **Sample**.
In cell B24 enter **Proportion marked (α_t)**.

In cell C24 enter **Size of marked population (M_t)**.
In cell D24 enter **Population estimate (N_t)**.
In cell E24 enter **Probability of survival (ϕ_t)**.

Under the Sample heading (first column), in cells A25 to A30, type the numbers **1** through **6** to indicate that there are six sampling sessions in the study. Next enter formulas for each of the remaining headings. Recall that the formula for proportion marked (α_t) is

$$\bar{\alpha} = \frac{m_t + 1}{n_t + 1}$$

This formula will be entered into the spreadsheet at cell B26. Type **0** in cell B25 because there are no marked animals in the first capture session yet. In cell B26, enter the formula for Proportion marked (α_t) as **=(D15 + 1)/(D17 + 1)**, and hit return. Notice that it is taking M_t, adding 1, and dividing that number by $n_t + 1$. This is the same formula for the proportion marked but with references to the cell containing the data instead of the actual number. Continue as follows:

In cell B27 enter the formula =**(E15 + 1)/(E17 + 1)** and hit return.
In cell B28 enter the formula =**(F15 + 1)/(F17 + 1)** and hit return.
In cell B29 enter the formula =**(G15 + 1)/(G17 + 1)** and hit return.
In cell B30 enter the formula =**(H15 + 1)/(H17 + 1)** and hit return.

It should now look like Table 9.10.

To enter the formula for the Size of marked population (M_t) in cell C25, click on cell C25, enter the formula =**((C18 + 1)*SUM(D4:H4))/(SUM(D5:H5)+1)+C15**, and hit return. Continue as follows:

In cell C26 enter =**((D18 + 1)*SUM(E5:H5))/ (SUM(E6:H6)+1)+D15** and hit return.
In cell C27 enter =**((E18 + 1)*SUM(F5:H6))/ (SUM(F7:H7)+1)+E15** and hit return.
In cell C28 enter =**((F18 + 1)*SUM(G5:H7))/ (SUM(G8:H8)+1)+F15** and hit return.
In cell C29 enter =**((G18 + 1)*SUM(H5:H8))/ (SUM(H9:H9)+1)+G15** and hit return.

Be sure to type the formulas exactly as listed, including double parentheses. It should now look like Table 9.11.

In cell D25, type **NA** (indicating "not available") because there is no estimate of population size with the data available from only one capture session. In cell D26 enter the formula =**C26/B26** and hit return. Notice that this is the size of the marked population (M_t) divided by proportion marked. Continue as follows:

In cell D27 enter =**C27/B27** and hit return.
In cell D28 enter =**C28/B28** and hit return.
In cell D29 enter =**C29/B29** and hit return.

It should now look like Table 9.12.

You could stop there and graph the changes in population size over time, or you could average the population size values to get a mean estimate of the population size. However, you might want to look at the probability of survival for each session. This can be calculated by clicking on cell E25 and entering the formula =**C26/ (C25+(C18−C15))** and hitting return. This is equivalent to

$$M_2/M_1 + (s_1 - m_1)$$

Table 9.10 Proportion of *Microtus* marked in each sample

Sample	Proportion marked (a_t)
1	0.000
2	0.375
3	0.326
4	0.283
5	0.404
6	0.563

Table 9.11 Proportion of *Microtus* marked in each sample and the size of the marked population

Sample	Proportion marked (a_t)	Size of marked population (M_t)
1	0.000	0.0
2	0.375	78.6
3	0.326	99.0
4	0.283	50.1
5	0.404	83.4
6	0.563	

Table 9.12 The estimated population size for the *Microtus* population on the grid

Sample	Proportion marked (a_t)	Size of marked population (M_t)	Population estimate (N_t)
1	0.000	0.0	NA
2	0.375	78.6	209.6
3	0.326	99.0	304.1
4	0.283	50.1	177.1
5	0.404	83.4	206.6
6	0.563		

Note: NA = not available.

Table 9.13 The probability of survival for the *Microtus* population on the grid

Sample	Proportion marked (a_t)	Size of marked population (M_t)	Population estimate (N_t)	Probability of survival (ϕ_t)
1	0.000	0.0	NA	0.407
2	0.375	78.6	209.6	0.955
3	0.326	99.0	304.1	0.392
4	0.283	50.1	177.1	0.946
5	0.404	83.4	206.6	0.000
6	0.563			

Note: NA = not available.

In cell E26 enter `=C27/(C26+(D18-D15))` and hit return.

In cell E27 enter `=C28/(C27+(E18-E15))` and hit return.

In cell E28 enter `=C29/(C28+(F18-F15))` and hit return.

It should now look like Table 9.13.

Save your spreadsheet. You can now enter your own data in the top table, and it will automatically recalculate the estimates in the lower table for you. (A detailed example of the Jolly-Seber method using the computer program JOLLY can also be found in Chapter 10 in this manual.)

Appendix

Table 9.14 Practice table for the Lincoln-Petersen method

Best guess at N	
M	
C	
R	
Lincoln-Petersen N	
95% confidence values	

Table 9.15 Practice table for the Schnabel method

Best guess at N	

Session	C_i	R_i	U_i	M_i
1				
2				
3				
4				
5				

$C_i M_i$	R_i

Schnabel N	
95% confidence values	

Table 9.16 Practice table for the Jolly-Seber method

	Time of last capture	Time of capture						
		1	2	3	4	5	6	7
	1							
	2							
	3							
	4							
	5							
	6							
Total marked (m_t)								
Total unmarked (u_t)								
Total captured (n_t)								
Total released (s_t)								

Sample	(a_t)	(M_t)	(N_t)
1			
2			
3			
4			
5			
Jolly-Seber N			

10 Using Software for Mark-Recapture Data

Time Required
- One 3- to 4-hour lab period

Learning Objectives
- Understand the basic concepts of closed and open capture models.
- Introduce the concept of model fitting.
- Use the program CAPTURE to understand capture models: M_0, M_t, M_b, and M_h.
- Understand how to compare and rank models.
- Use RStudio to run and understand closed and open capture models.

Equipment Required
- Computers with access to the internet

Background

Recall from Chapter 9 that a closed population is one that has no additions or deletions of animals from one sampling period to the next. In other words, there cannot be any births or deaths, nor any movement of individuals into or out of the study area during the sampling periods. A study that is run over a short period (e.g., consecutive days) will likely meet these assumptions. In contrast, a mark-recapture study where data are collected over several months or years will likely violate the assumptions of a closed population.

In studies where animals are abundant or the survey area is large, the resulting data set may be too large to allow for accurately estimating population size (N) by hand (as was done in Chapter 9). Fortunately, there are a number of software packages that can be used to analyze these more complex data sets. Most of these involve sophisticated statistical techniques, such as "Maximum Likelihood Estimation" that are beyond the scope of this discussion (but see Williams et al., 2001). We will use online versions of CAPTURE software and the R package Rcapture to analyze several sample populations. Before we dive into the analysis, it is important to understand a few basic concepts.

Capture Probability and Encounter Histories

Assume we complete a three-day mark-recapture study. On the first day, we capture, mark, and release 55 rodents on our study grid (each animal receives a unique mark). On days two and three we capture, record marks, and release the animals without adding any new marks to the unmarked individuals (so for simplicity of this discussion we always have 55 marked animals in the population). The basic model is shown in Figure 10.1.

In Figure 10.1 we have three sampling days (session 1 to 3) denoted by boxes. Time runs from left to right. The arrows between sessions denote the probability that an animal marked in session 1 will survive until sampling session 2 (ϕ_1) and that, having survived until session 2, the animal will survive until session 3 (ϕ_2). The probabilities below sessions 2 and 3 are the probability of encountering a marked animal (p_1 and p_2). Thus, the probability of encountering any animal at a session is determined by both ϕ and p.

Recall that we only marked animals in the first session. Every time we encounter a marked animal we record the encounter with a 1. If that animal was not seen in a session, it is recorded as a 0. We can create an encounter history for each of the 55 animals originally marked (Figure 10.2). In Figure 10.2, the top encounter history would mean that this individual was marked in session 1 and recaptured (encountered) again in both session 2 and session 3. To be encountered, it must also have survived the intervals between sessions 1 and 2 and sessions 2 and 3. Thus, the probability of recording this

Figure 10.1
Encounter history and probability for each session.

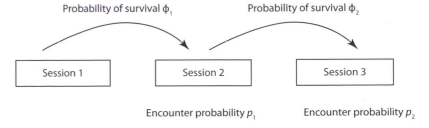

Probability of survival ϕ_1 · Probability of survival ϕ_2

Session 1 · Session 2 · Session 3

Encounter probability p_1 · Encounter probability p_2

Encounter history	Probability
1 1 1	$\phi_1\ p_2\ \phi_2\ p_3$
1 1 0	$\phi_1\ p_2\ (1 - \phi_2\ p_3)$
1 0 1	$\phi_1\ (1 - p_2)\ \phi_2\ p_3$

Figure 10.2 Encounter history for three animals along with the probability of that encounter history.

particular encounter history (e.g., 111) is equal to $\phi_1\ p_2\ \phi_2\ p_3$.

The second animal in Figure 10.2 has an encounter history of 110. This animal was marked in session 1 and recaptured (encountered) in session 2 but not encountered in session 3. The third animal has an encounter history of 101, corresponding to being marked in session 1, not captured in session 2, but recaptured in session 3. Obviously, even though it was not captured in session 2, it did survive that interval. There is one encounter history that is possible but not shown. What is it, and what would its probability be? Could this study yield an encounter history of 011?

Obviously, with 55 marked animals, several animals will share a particular encounter history. We could list each of the 55 animals separately in a long column. Alternatively, because we know there are only four possible encounter histories, we can add this information to our encounter history data as follows:

    111   7
    110   20
    101   15
    100   13

This set of encounter histories says that 7 animals have an encounter history of 111, 20 animals have an encounter history of 110, and so on.

To use the software to calculate the probabilities and maximum likelihood estimates, we need to put our mark-recapture data into an encounter history format that the software can understand. The basic data format for encounter histories for each animal caught over the course of the study is a matrix whose columns represent trapping sessions (time periods) and whose rows represent animals. The matrix entries are either "1" (denoting a capture) or "0" (denoting not captured). Consider the set of encounter histories in Table 10.1.

Notice that in Table 10.1 it is possible to mark new animals as they are encountered in the study (e.g., a 0 in the first column). How many individuals are represented in population A? How many in population B?

Capture Models

As mentioned, complex statistical models have been developed to model capture probabilities and estimate pop-

Table 10.1 Sample mark-recapture encounter histories

Population A	Population B
101111	1011 2
001010	1100 10
010000	1101 1
000001	1001 4
110010	0101 3
100011	1011 12
111000	1010 8
100010	1000 10

ulation size (N). If there are only two sampling periods, however, these models do not generally apply, and it is best to use the Lincoln-Petersen method described in Chapter 9. When the number of sampling sessions is greater than two, then a number of models are available. We will look at several models that vary in certain ways:

> M_0—Neither behavioral nor temporal variation nor capture heterogeneity
> M_b—Behavioral response only
> M_t—Temporal variation only
> M_h—Individual capture heterogeneity
> M_{bh}—Behavioral response and capture heterogeneity

Let us look at each model a bit more closely. The first model, M_0, assumes that there is no variation in capture probability over time (between sessions) and no change in the animal's behavior toward the traps (e.g., no trap-shy or trap-happy animals). This is the simplest model because if all encounter and survival probabilities are the same, then the model looks like that in Figure 10.3.

Obviously, this is not always realistic. The behavioral model (M_b) allows changes in the behavioral response between sessions, such as a change in capture probability associated with previous capture history. For example, marked and unmarked animals can now have different encounter (capture) probabilities. This can occur when animals become trap-happy (i.e., an increased probability of being captured after the initial capture) or trap-shy (a decreased capture probability in subsequent sessions).

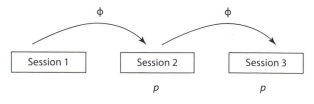

Figure 10.3 Probability history for the M_0 model.

The temporal model (M_t) assumes that each animal has the same capture probability for each trapping session, but capture probabilities vary from one session to the next. In the individual capture heterogeneity model (M_h) there is no variation in temporal or capture probabilities, but each animal may have a different capture probability. Finally, the M_{bh} model has two sources of variation: behavioral variation and potentially different capture probabilities between animals (details and mathematical equations for each can be found in Williams et al., 2001).

Exercise 1: Using the Program CAPTURE

The CAPTURE program calculates estimates of capture probability and population size for closed population mark-recapture data (Rexstad and Burham, 1991; White et al., 1978). CAPTURE generates output for

- models M_0, M_b, M_t, M_h, and M_{bh} (along with several other models),
- goodness of fit and model comparison statistics,
- estimates of abundance,
- capture probability and standard error (SE) of the estimates under each model, and
- a test of the closure assumption.

CAPTURE can be downloaded as a stand-alone software package for Windows PCs from the Patuxent Software Site (http://www.mbr-pwrc.usgs.gov/software .html), or you can use the convenient online version at https://www.mbr-pwrc.usgs.gov/software/capture.html.

You will use the online version in this exercise. To get started, you'll need some mark-recapture data in the format readable by CAPTURE. Type (or cut and paste) in the data in Appendix A.

The top three lines are:

TASK READ CAPTURES OCCASIONS=6 × MATRIX
FORMAT='(A1, 1x, 6F1.0)'
READ INPUT DATA

The first line tells the program CAPTURE to read a set of data in a capture matrix with 6 encounter sessions (occasions = 6). There are four ways data can be entered, but you will only learn the matrix format here (for details on the others, consult White et al., 1978). The second line is a format statement that tells CAPTURE that the data have a label (each animal is given a unique letter), followed by a space, followed by a string of 6 numbers. In this case the 6 numbers are a string of "1s" and "0s," where a 1 represents an encounter (capture) and a 0 represents no encounter in that session. The third line tells CAPTURE to input the data that follow. Below the three commands is the data matrix itself, beginning with animal "a" with its encounter history of "100010" and ending with animal "z" and its encounter history "001111."

```
a 100010
b 100000
c 100010
. . .
z 001111
```

The data matrix is followed by more command lines (tasks) that tell CAPTURE what you want it to do with the data. In this example, we are asking CAPTURE to test the models and determine the most appropriate model to fit to this data set, and to estimate the population size. The commands are:
TASK MODEL SELECTION
TASK POPULATION ESTIMATE ALL

The first **TASK MODEL SELECTION** command produces a series of hypothesis tests that help determine which of the models provides the best estimate. The second command provides estimates of population size for each model.

1. Type (or cut and paste) the lines in Appendix A into the white data entry box in the online version of the program CAPTURE.

2. Click on the Perform Analysis button at the bottom of the page. A new window opens with a lot of useful information. It is best to print this output (or cut and paste it into a text file) now so that you can follow along with the description of the output below.

Data Output in Capture

This simple analysis yields about 15 pages of output. What does it all mean? The first information you see in the output file is a summary of what the output file contains. Sometimes you'll see an input error (unidentified task) listed here, but usually this is due to an extra line and can probably be ignored. If you get a lot of input errors, it means you typed in the data incorrectly.

In this analysis, the output file should begin with
CAPTURE output . . .
Mark-recapture population and density
estimation program
Page 1
  Program version of 16 May 1994
Input and Errors Listing
Input---task read captures
occasions=6 × matrix
** Warning ** captures= 6.00000 assumed.
Input---format='(A1, 1x, 6F1.0)'
Input---read input data
  Summary of captures read
    Number of trapping occasions 6
    Number of animals captured 26
    Maximum x grid coordinate 1.0
    Maximum y grid coordinate 1.0

You should confirm that this is appropriate for the data set you used. Remember, the program does what you tell it to; as the saying goes, garbage in, garbage out. It is the user's responsibility to ensure that the analysis is performed correctly. These lines are followed by more information. For our purposes, skip down to page 2 and find the lines:

```
Model selection procedure. See this
section of the Monograph for details.
Occasion j=1 2 3 4 5 6
Animals caught n(j)=23 19 11 0 0 0
Total caught M(j)=0 23 25 25 25 25
Newly caught u(j)=23 2 0 0 0 0
Frequencies f(j)=4 14 7 0 0 0
```

Notice that in this table the number of animals caught over the six sessions was fairly constant. This is consistent with a closed population. If the number of captures had increased sharply in sessions 5 and 6, then it might indicate that individuals immigrated onto the trapping grid toward the end of the study. Conversely, if the number of captures dropped off sharply toward the end of the study, it might indicate that some individuals either left the study area or died. In either case, the assumption of a closed population is violated. In this example, you can assume that the population is closed.

Following this summary information about the various models, skip down to the table for **Model selection criteria** on page 3 of the output. It should look like Table 10.2.

The values for each model in this table can be roughly interpreted as the probability that the model is the "correct" model. Model M_{tb}, having the highest value (1.00), appears to be the best model. However, use caution in accepting the weight given in this table. For example, when data sets are small, M_0 will probably be weighted 1.00 and therefore the "best" model when in fact it is not.

Below this table in the output report, you should find output for each model. Find each of the models, and create a summary table of their estimates for population size, standard error, approximate 95% confidence intervals, and profile likelihood intervals. Write a short summary of the results, emphasizing the model you chose and explaining your rationale.

Table 10.2 Summary table for the CAPTURE data in Appendix A

Model	M_0	M_h	M_b	M_{bh}	M_t	M_{th}	M_{tb}	M_{tbh}
Criteria	0.17	0.06	0.70	0.10	0.00	0.34	1.00	0.51

Exercise 2: Tigers in India

Repeat this analysis using the second set of data (in Appendix B). These data are from a survey of tigers in Kanha National Park in Madhya Pradesh state in central India (Karanth et al., 2004; data and exercise are modified from http://bcss.org.my/stats/ClosedCap_tigers.htm). Instead of using livetrapping to mark and release the tigers, the researchers used trail cameras to "trap" tigers, and individuals were recognized in subsequent photos (captures) by their unique pattern of stripes (marks). A total of 803 camera trap-days over a three-month period were collected in this study (see Chapter 12). During this period, 26 individual tigers were identified. Several tigers were "captured" on photos two or three times, others only once, allowing the researchers to generate a capture history for each tiger (Appendix B). Because of the short study period (three months) relative to the lifespan and gestation period of tigers, it is reasonable to assume the population was closed.

1. Run CAPTURE with the data from Appendix B (follow the directions for Exercise 1).

2. CAPTURE output starts with a summary of the input data. Check this to be sure it is accurate.

3. CAPTURE carries out model tests. Look them over, remembering that a large probability value means that the data fit the model well. The first three heterogeneity tests all compare M_0 against other models.

4. CAPTURE combines the results of the goodness-of-fit tests and uses an algorithm developed from simulated data sets to calculate the most appropriate model. Recall that the most appropriate model has a score of 1.00. The model tests for the tiger data are summarized below in Table 10.3.

Answer the following questions:

- What is the most appropriate model?
- What are the population size estimates, standard errors, and approximate 95% confidence intervals for this model?
- What model would you choose that would be more robust than the one considered in question 1?

Table 10.3 Summary table for the tiger data in Appendix B

Model	M_0	M_h	M_b	M_{bh}	M_t	M_{th}	M_{tb}	M_{tbh}
Criteria	0.69	0.69	0.59	0.52	0.00	0.24	1.00	0.58

- What are the values for population size and 95% confidence intervals for the second-best model?
- The M_0 model assumes that the capture probabilities for all tigers are equal. Is that a reasonable assumption? Note: This would mean that all tigers behave the same way.
- What is your best estimate of the number of tigers in Kanha National Park, with an approximate 95% confidence interval?
- Kanha National Park is 282 km² in area. What is the density of tigers in the park, given as tigers per 100 km²?

Exercise 3: Mark-Recapture Sampling Using Rcapture

This exercise assumes that you or your instructor has some familiarity with using RStudio, a graphical interface for R. The workflow is to open RStudio, get the Rcapture package (Baillargeon and Rivest, 2007), import data from Excel, and estimate abundance using both closed and open population models. The Excel files can be downloaded from the Johns Hopkins University Press website for this manual (jhupbooks.press.jhu.edu/content/mammalogy-techniques-lab-manual).

The R package Rcapture is used to analyze capture-recapture experiments. The data for analysis consist of encounter histories over several capture sessions. Recall from Chapter 9 that encounter histories are groups of zeros and ones where one is a capture and zero is no capture in a given session. Rcapture can fit three types of log-linear models. However, we will only be concerned with a closed population model and an open population model. In Chapter 9 we learned that a closed population model assumes no births or deaths and no immigration or emigration during the experiment. In an open population, immigration/emigration and births/deaths occur between sampling periods. The open population models give an estimate of N but also provide survival rates using the Cormack-Jolly-Seber model.

In the closed population model Rcapture fits several models to the data. The goal is to choose the best model to estimate population size (N). The closed population function in Rcapture is the command "closed," which fits the models M_0, M_t, M_h, M_{th}, M_b, and M_{bh} (discussed above). The command "closed" generates deviances, degrees of freedom, and Akaike Information Criteria (AIC) values. From these values, the model with the lowest AIC is typically the best-fit model.

In our exercise, we will fit closed population models to data on snowshoe hares (*Lepus americanus*) originally described in Agresti (1994) and Cormack (1985). These data are already included in the Rcapture package.

1. Open RStudio and load the Rcapture package by entering the three commands shown below in Courier font (do not include the >; this is the prompt).

```
install.packages("Rcapture")
require(Rcapture)
library(Rcapture)
```

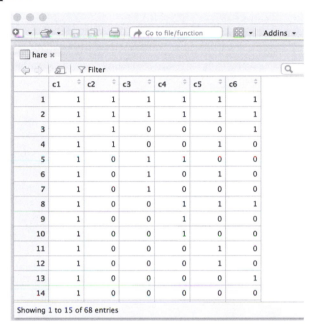

Figure 10.4 The Console window in RStudio showing the data in the hare file. The six capture sessions are listed across the top for each of the 68 hares in the data set.

2. To attach the data for the snowshoe hares, enter the command

```
data(hare)
View(hare)
```

You will now see the raw data in the window above the Console window. It should look similar to Figure 10.4. The data are for 68 hares (rows) in a study with six capture sessions (columns 1–6).

3. First we will look at some descriptive statistics for the data by entering the command:

```
descriptive(hare)
```

The descriptive command prints out the number of captured units (hares) as 68 as well as a set of frequency data for the number of hares captured i times (f_i). Take a look at row $i=6$ and you will see f_i equals 2, meaning that only 2 of the 68 hares were captured in all 6 trapping sessions. You will also see statistics for u_i, v_i, and n_i. Can you explain what they mean? Can n_i for session 6 ever be less than for session 2?

4. Now we will estimate hare abundance using the closed population models using the command:

```
closedp(hare)
```

Number of captured units: 68
Abundance estimations and model fits:

	abundance	stderr	deviance	df	AIC	BIC	infoFit
M0	75.4	3.5	68.516	61	154.707	159.146	OK
Mt	75.1	3.4	58.314	56	154.505	170.041	OK
Mh Chao (LB)	79.8	6.4	58.023	58	150.214	161.311	OK
Mh Poisson2	81.5	5.7	59.107	60	147.298	153.956	OK
Mh Darroch	90.4	11.6	61.600	60	149.791	156.449	OK
Mh Gamma3.5	100.6	21.7	62.771	60	150.961	157.619	OK
Mth Chao (LB)	79.6	6.3	47.115	52	151.305	175.720	OK
Mth Poisson2	81.1	5.6	48.137	55	146.327	164.083	OK
Mth Darroch	90.5	11.7	50.706	55	148.896	166.652	OK
Mth Gamma3.5	101.6	22.4	51.956	55	150.147	167.903	OK
Mb	81.1	8.3	67.027	60	155.217	161.876	OK
Mbh	74.2	14.6	63.257	59	153.447	162.325	OK

Again, Rcapture prints out a table of data for each model. The table should look like the one above.

The models appear in rows with the name of the model on the far left (M_0, M_t, etc.). The second column is the abundance estimate for each model. These estimates range from 74.2 to 101.6 snowshoe hares in the study population. The AIC column shows that the model with the lowest AIC value is M_{th} Poisson2 with 146.327. This is usually the best model (assuming no heterogeneity), and it suggests that the population size is 81.1 hares.

5. Now we will look at the snowshoe hare data as an open population. Open population models are used when the capture sessions are far apart (weeks to months) and animals move in and out of the population during the experiment. Rcapture uses the command "openp" to fit the Cormack-Jolly-Seber and the Jolly-Seber models. Begin by entering the commands

```
op <- openp(hare, dfreq = TRUE)
op$model.fit[1, ]
```

This runs the openp function on the hare data and calls the new dataframe "op" (for open). The output shows a deviance of 17.36 and an AIC of 80.68. Next print out the estimated population size (N) for an open population by entering the command:

```
op$N
```

This prints out the abundance estimates for each period. Notice they are lower because mortality and emigration are allowed.

	estimate	stderr
period 1	NA	NA
period 2	20.91656	6.215693
period 3	24.27497	3.594807
period 4	27.63636	7.172868
period 5	NA	NA

6. For an estimate of population size over all six capture sessions, use the command:

```
op$Ntot
```

This yields an estimate of 30.5 hares in the population. We can also get an estimate of the number of births at each session using the command

```
op$birth
```

There is much more to explore in Rcapture. If you are interested, look at the Rcapture manual for additional models and examples.

Appendix A

```
TASK READ CAPTURES OCCASIONS=6 × MATRIX
FORMAT='(A1, 1x, 6F1.0)'
READ INPUT DATA
a 100010
b 100000
c 100010
d 110001
e 111000
f 110100
g 101100
h 111110
i 101110
j 110111
k 101100
l 010100
m 010010
n 011001
o 011100
p 011111
q 001101
r 011011
s 010111
t 010100
u 010001
v 001000
w 010011
x 001001
y 001110
z 001111
task model selection
task population estimate ALL
```

Appendix B

```
TASK READ CAPTURES OCCASIONS=10 × MATRIX
FORMAT='(A1, 1x, 10F1.0)'
READ INPUT DATA
a 1001000110
b 1000000101
c 1101000011
d 0110000111
e 0101000001
f 0100000000
g 0010011000
h 0011000000
i 0001000011
j 0001100110
k 0001001001
l 0000100000
m 0001000000
n 1001000000
o 0001000000
p 0001100000
q 0001001000
r 0000110000
s 0000100000
t 0000010000
u 1000101000
v 0100000000
w 0001000001
x 0000010000
y 0000000001
z 0000100000
task closure test
task model selection
task population estimate ALL
```

11   Transects: Using Distance Sampling

Time Required

- One 3- to 4-hour lab period for conducting the transect field study and another 3- to 4-hour lab period for the in-lab data analysis using Distance software.

Learning Objectives

- Learn how to set up a line transect study.
- Understand how to record data for a line transect study.
- Understand the difference between indirect and direct counts.
- Use deer pellet counts to estimate population density.
- Use the RStudio program to analyze both indirect and direct transect data.
- Become familiar with the R statistical environment.

Equipment Required

- 30-meter tape measure or topofil and hip chain
- Fluorescent vinyl flagging tape
- Marker pens
- Maps of the study sites
- Handheld GPS
- Altimeter and clinometer
- Optical range finder or survey laser binoculars
- Data sheets and clipboard
- Field notebook
- Camera (optional)
- Access to computers connected to the internet

Background

Sampling mammal populations in large areas (e.g., national parks or reserves) is difficult because it is impossible to count every animal in the area. Researchers must use a subsample of the entire population. Using quadrants or grids is not feasible for large mobile mammals, so researchers typically use line transects for these types of surveys. Transect lines are established in the area of interest, and each line is methodically and uniformly searched for the target species (Figure 11.1). If you could be assured of counting every animal along a transect of fixed width, then the transect line simply acts as a very thin rectangular quadrant (Krebs, 1999). However, it is unlikely that an observer will be able to count every animal along a transect (e.g., undetected animals, see Buckland et al., 1993). In such cases, we need to know the probability of detection before we can estimate density.

When an animal is observed, three measures are taken:

r_i = sighting distance from the observer to animal i

θ_i = sighting angle between the animal, the observer, and the transect line

d_i = perpendicular distance from animal i to the transect line (calculated as $d = r \sin \theta_i$)

Notice that in Figure 11.1 not all individuals are detected as the observer moves along the transect. In fact, the probability of detecting an animal decreases as its distance from the line increases. Thus, we need a way of estimating the detection function in order to have an accurate measure of the final density of animals. The de-

Figure 11.1 Illustration of a line transect and the variables required to estimate density.

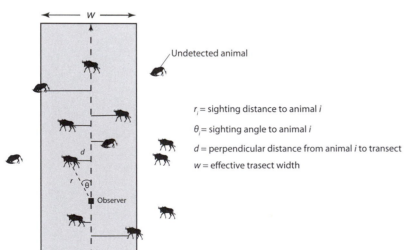

Undetected animal

r_i = sighting distance to animal i

θ_i = sighting angle to animal i

d = perpendicular distance from animal i to transect

w = effective trasect width

Observer

Transect line

tection function, $g(x)$ = probability of observing an animal at x distance from the transect line.

We then estimate density (D) as

$$\hat{D} = \frac{n}{2La}$$

where n = number of animals observed, L = total length of transect, and $a = 1/2$ the effective transect width (i.e., the area under the detection function).

The basic problem in estimating density is estimating the parameter a, and there are numerous methods for estimating a in the literature (see Krebs, 1999). In exercise 3, we will use R package Rdistance to estimate a and provide an estimate of the total density of animals.

Indirect Data

In some cases, it is not practical to survey animals by direct observation. For example, some species are secretive and rarely seen, and others range over vast areas. In these cases, it may be possible to use indirect evidence of the animal's presence to estimate density from transect surveys. Typically, this involves using signs such as feces (pellet piles of deer or dung piles of elephants) or nests (of chimpanzee and orangutans). Such indirect methods require knowing (1) the defecation rate—i.e., the number of dung piles produced per animal per day—or nest-building rate, and (2) the decay rate (how long it takes for the dung or the nest to disappear in the environment).

Dung count estimates have several limitations (Plumptre, 2000). First, it is often difficult to determine the species responsible for the dung pile. Second, defecation rates vary with diet. Third, weather conditions (and dung beetle activity) greatly affect the decay rate of dung piles. Finally, dung piles are often used to mark territory and are not randomly distributed (leading to statistical problems). Nevertheless, dung counts may be the best option for large mammal surveys in forested areas. Dung counts have been used for surveying forest elephants (Barnes and Jensen, 1987), buffalo, and forest duikers. For elephants, dung count density estimates appear to correlate well with those from other methods (Barnes, 2001).

Field Procedures

To survey large areas quickly or with few personnel (two or three surveyors), there are two options: a reconnaissance survey or transects. Reconnaissance surveys are quick-and-dirty assessments that do not provide sufficient data for estimating density, but they may be a first step toward a more detailed census of the area. Unlike transects, reconnaissance surveys follow a winding path of "least resistance" through the habitat to cover as much ground as possible. In contrast, line or strip transects

provide data for estimating relative abundance or population density. Transect data can be collected for direct observation of animals or indirect observation via dung or other signs (see exercises below).

The following are some general guidelines for conducting most line transect surveys (see Anderson et al., 1979):

- If nocturnal species are present, surveys should be carried out by night and day.
- New transects should be located to sample different vegetation types and/or levels of disturbance.
- Transect(s) should be straight and well marked (a series of straight line segments may be more appropriate for some areas).
- All animals directly on the transect line must be seen (i.e., with probability equal to one).
- All distances and angles must be accurately measured.
- At least 40 animals should be seen ($n > 40$), and 60–80 are preferred. Therefore, transect surveys can only generate population estimates for species that are relatively abundant and visible.
- A pilot survey is recommended to aid in planning the survey design. Often, a simple visit to the area to be surveyed, along with basic biological information about the animal and its habits and habitat, will be sufficient to design an adequate survey to estimate density.
- Avoid transects running along roads, streams, or other potentially biasing features.

Transects should cover the main habitats in the approximate proportions as they occur in the area and should not cross each other. Transects should be positioned at least 300 meters apart. New paths should be cut before surveying to avoid frightening animals during the survey and to avoid the use of animal trails, logging roads, and other such features, which would bias the survey. Ideally, transects are cut by first selecting a starting point and direction using a random number table. However, if the goal is to sample all habitats in the area, this approach may not be possible. Each transect should be kept straight using a compass or GPS device to maintain a constant heading. Record and map any streams, buildings, and roads as you go (using a GPS unit). It is good practice to mark the distance along the transect every 50 meters by writing the distance on flagging attached to a tree along the transect line (remove all flagging at the end of the survey period if the transect will not be reused).

Daytime surveys should be conducted when animals are active, from just after dawn to about 11:00 a.m. If this is not practical, afternoon surveys can be carried out between 15:00 and 17:30 hours. To avoid bias, do not conduct surveys during the rain. Transect surveyors should walk slowly and quietly, scanning from side to side for movements and listening for sounds. An average speed of 1 kilometer per hour, with a pause every 100 meters or so is common. This does not include the time it will take to collect data if an animal is sighted. Before beginning each transect walk, record weather and other general information on a transect survey form. When an animal is sighted, record the time, species identity, group size, group spread, and sighting location along the transect onto the data sheet. Be sure to record subsidiary information such as activity, diet, height, age, and sex of animals sighted, mixed-species associations, and vegetation features. Record sighting distances (SDs) and angles using a pre-calibrated range finder and a sighting compass. Convert these to perpendicular distances r (sin θ_i), and record this on the data sheet. (Note: It may be easier to memorize the exact location of the animal, walk to the point perpendicular to the transect, and measure from this location). Finally, record any other observations, such as the number of males and females in the group, the number of juveniles, or any behaviors on your data sheets or field notebook.

Indirect Transect Surveys

Transect surveys are best when the species are highly visible and the habitat is relatively open. When species are secretive or the vegetation is dense, accumulating enough observations is difficult, and you may need to use an indirect approach. Indirect sampling uses dung or fecal pellet counts as a surrogate for estimating population density. Fecal counts have been used to estimate populations of deer (Marques et al., 2001; Mooty et al., 1984; Batcheler, 1975), moose (Härkönen and Heikkilä, 1999), and rabbits (Rouco et al., 2016). Nests have been used as an indirect measure of chimp populations (Kouakou et al., 2009).

One complicating factor involved in fecal counts is that fecal pellets or dung piles persist for days or weeks depending on the weather and season. A second complication is that animals produce dung/pellets at different rates. Thus, both decay rate and defecation rate must be known.

In 2001, the Wildlife Conservation Society-Indonesia Program estimated the elephant population of Bukit Barisan Selatan National Park (BBSNP) in Sumatra, Indonesia, from dung surveys (Hedges et al., 2005). The defecation rate (dung production rate) was estimated by observing 12 captive (but free-ranging) elephants in nearby Way Kambas National Park. Dung decay rates (disappearance time) were based on monitoring 1,302 dung piles in BBSNP for 18 months prior to the transect survey. Finally, dung pile density in BBSNP was estimated by line transect surveys.

If each elephant produces p dung piles per day (defecation rate) and dung piles remain visible for t days (decay rate), then dung density (S) will be

$$S = D \times p \times t$$

where D is the density of elephants. If you can estimate p and t, and you run transects to determine S (dung density), then you can estimate the density (D) of elephants without actually seeing them. Preliminary surveys for forest elephants in BBSNP revealed 1,313 elephant dung piles along transect lines totaling 73.6 kilometers. The estimated defecation rate was 18.15 dung piles per 24 hours with a standard error of 2.53. Dung piles remained visible for 305.36 days with a standard error of 7.33. Therefore, elephant density = $1,313/(18.15 \times 305.36) = 0.23$ elephants per square kilometer, or approximately 498 elephants in the park.

Exercise 1: Conducting Deer Pellet Transect Surveys

There are two main methods for counting fecal pellets or dung piles. The first is the fecal accumulation rate (FAR) method. It requires two survey visits per transect. During the first visit, strip transects are marked out and all dung/pellets are removed from the transect. On the second visit, all dung or pellet piles are counted. The second visit must be close enough to the first visit so that animals will have produced fresh dung but not so long after that the dung will have decayed. The second method, the fecal standing crop (FSC) method, requires only one visit to each transect. This method requires that the decay rate of fecal material be known in advance for each habitat type. The exercise below uses the FAR method.

1. Your instructor will provide you with a map of an appropriate study area. Break down the study area into different habitat types using imagery from Google Earth.

2. Establish a series of strip transects through each habitat type. Each transect should be 500 meters long by 2 meters wide (with an area totaling 1,000 square meters).

3. On the first visit, begin at one end of the transect and slowly move down the transect. It can be helpful to carry a 2-meter rod held horizontally to better delineate the edge of the transect. Count all pellet groups with any part of the group inside the 2-meter transect boundary. This count can be used for the FSC method later.

4. Remove all pellet groups within the transect (wear gloves and deposit them into a large garbage bag).

5. When you encounter a pellet group that is fresh, move it 1 meter from the transect and mark it with a flag. These groups can be used to determine decay rate.

6. Return to the transects two to three weeks later, and count all pellet groups within the borders of the

transect (using the 2-meter rod as a guide). For each pellet group encountered, record its distance along the transect in 50-meter intervals (e.g., "1 pellet group at 100-150m interval"). Also, note the degree of decay in the marked pellet groups.

7. For the FAR method, look up the defecation rate for your deer species. Rogers (1987) estimated 34 pellet groups per individual per day for wild white-tailed deer. Sawyer and colleagues (1990) estimated 26.6 pellet groups/individual/day for a captive population in Georgia. However, Smith (1964) reported an average pellet group per deer per day of 13-14 for captive mule deer. Search the literature for an estimate that is from the same state or region and the same time of year. Alternatively, use a range of estimates in your analysis.

8. Assuming the marked pellet groups were still present during your second visit, estimate the density of deer in the study area using the following formula:

$$\text{Deer per km}^2 = \frac{\text{\# pellet groups per km}^2}{\text{days between visits} \times \text{defecation rate}}$$

For example, suppose you sampled 4 transects that were each 100 meters long (with a total area of 800 square meters), recorded 52 pellet groups, and estimated a defecation rate of 26 for your species and locality. Your equation would be

$$\text{Deer per km}^2 = \frac{0.065 \times 1{,}000{,}000}{14 \times 26} = \frac{65{,}000}{364}$$
$$= 178 \text{ deer / km}^2$$

The numerator is the number of pellet groups per square meter (52/800) multiplied by 1 million to convert from meters to kilometers.

9. If you know the decay rate of the pellet groups, you can also use the FSC method:

$$\text{Deer per km}^2 = \frac{\text{\# pellet groups per km}^2}{\text{defecation rate} \times \text{decay rate}}$$

Otto (2014) found the decay rate for white-tailed deer in Ohio to be 15 days. Using this value in our example above yields

$$\text{Deer per km}^2 = \frac{65{,}000}{26 \times 15} = 167 \text{ deer / km}^2$$

Exercise 2: Dung Counts Using PELLET

In 2014, Salvador Mandujano developed an Excel spreadsheet approach to analyzing pellet/dung count data called PELLET. This spreadsheet calculates both FAR and FSC for large data sets. The spreadsheet is available under the Creative Commons Attribution 3.0 license (http://creativecommons.org/licenses/by/3.0/us/). It can be downloaded from the web at http://www1 .inecol.edu.mx/cv/CV_pdf/mandujano/PELLET _English.rar.

Download and open PELLET. The spreadsheet opens with a large data set for white-tailed deer in Oaxaca, Mexico, from 2013. To understand the spreadsheet, first notice that there are actually four sheets in the workbook. The sheets are ST-FSC, ST-FAR, TCP-FSC, and TCP-FAR. Locate them in the tabs along the bottom left of the spreadsheet. FAR and FSC methods are described in exercise 1 above. ST refers to a strip transect, and TCP refers to a center point transect (described in Mandujano, 2014).

Across the top of the spreadsheet (Figure 11.2) are the transects (T1-T8) and their dates. The transects in this study were each 500 meters long divided into 50-meter subtransects. For example, in transect T1 no pellet groups were observed in the first 50 meters, but three pellet groups were encountered in the next 50 meters of transect. Transect T1 had 11 pellet groups total. Below the transect are four cells containing summary information for each transect.

PELLET includes several estimates of defecation rate and deposit time. Those values are found in the large table below the transect data. To the far right are two summary tables. The top shows the negative binomial distribution. The bottom summary table contains the mean density of 4.75 deer per square kilometer and the standard deviation (2.4) as well as the confidence intervals (1.9 to 13.9). Click on the cell containing the 4.75 value to reveal its formula = AVERAGE(C19:J36). This means it is the average density of all the density estimates in cells C19 through J36. Explore other cells to reveal their formulas (be careful not to change them).

1. Make a copy of the ST-FAR sheet by clicking on the ST-FAR tab and going to Edit > Sheet > Move or Copy Sheet in the menu at the top. A popup window appears asking where you want to put the new sheet. Click "(move to end)", then click the box next to "Create a copy," then click "OK". This puts a new copy of that sheet in the tabs at the bottom. Double click on the new tab, and edit the name of the sheet with the name of your study site.

2. Clear the entries in the newly created sheet by selecting them and typing in your own data. Be sure to edit the site name in cells A1 and A2 as well as the dates under each transect.

3. Delete any unnecessary transect columns. For example, if you only have 10 transects, click on the M

Figure 11.2 The PELLET spreadsheet for estimating deer density from pellet-group counts on strip transects.

		T1	T2	T3	T4	T5	T6	T7	T8				
Sampling date of each transect →		01/05/2013	01/05/2013	08/05/2013	07/05/2013	09/05/2013	09/05/2013	01/05/2013	09/05/2013				
0-50 m		0	1	0	0	0	0	0	0				
51-100 m		3	3	7	3	8	3	3	3				
101-150 m		0	1	0	0	0	0	0	0				
151-200 m		7	1	7	7	7	9	7	7				
201-250 m		1	1	5	1	1	1	1	1				
251-300 m		0	0	0	0	0	0	0	0				
301-350 m		0	1	4	0	0	0	0	0				
351-400 m		0	0	0	0	0	0	0	0				
401-450 m		0	0	0	0	0	0	0	0				
451-500 m		0	1	3	0	0	0	0	0				
Number pellet-group / transect		11	9	26	11	16	13	11	11	108			
mean pellet-group / parcel / transect		1.1	0.9	2.6	1.1	1.6	1.3	1.1	1.1				
MAX deposit days of pellet-groups		166	166	173	172	174	174	166	174				
MIN deposit days of pellet-groups		150	150	157	156	158	158	150	158				

Negative Binomial distribution calculation

No. Transects	8
No. Parcels (n)	80
Total Pellet-groups	108
Mean (X) pellet-groups/ trans	13.500
Mean (X) pellet-groups/ parce	1.350
Variance (S²) pellet-groups/ p	0.297
NegBin coefficient (k)	0.061

Possible dates of deposited pellet-groups	Defecation rate	Densities estimations (deer/km2) per transect, defecation rate, deposite tim								Density	±SD
15/11/2012 MAX	12.2	5.4	4.4	12.3	5.2	7.5	6.1	5.4	5.2	6.5	2.5
		5.7	4.7	12.6	5.5	7.8	6.4	5.7	5.5	6.8	2.5
		5.1	4.1	12.0	5.0	7.3	5.8	5.1	4.9	6.2	2.5
	19.3	3.4	2.8	7.8	3.3	4.8	3.9	3.4	3.3	4.1	1.6
		3.6	3.0	8.0	3.5	4.9	4.1	3.6	3.5	4.3	1.6
		3.2	2.6	7.6	3.1	4.6	3.7	3.2	3.1	3.9	1.6
	26.4	2.5	2.1	5.7	2.4	3.5	2.8	2.5	2.4	3.0	1.2
		2.6	2.2	5.8	2.6	3.6	3.0	2.6	2.5	3.1	1.2
		2.4	1.9	5.6	2.3	3.4	2.7	2.4	2.3	2.9	12
01/12/2012 MIN	12.2	6.0	4.9	13.6	5.8	8.3	6.7	6.0	5.7	7.1	2.8
		6.3	5.3	13.9	6.1	8.6	7.1	6.3	6.0	7.5	2.8
		5.7	4.6	13.3	5.5	8.0	6.4	5.7	5.4	6.8	2.8
	19.3	3.8	3.1	8.6	3.7	5.2	4.3	3.8	3.6	4.5	1.8
		4.0	3.3	8.8	3.9	5.4	4.5	4.0	3.8	4.7	1.8
		3.6	2.9	8.4	3.5	5.0	4.1	3.6	3.4	4.3	1.8
	26.4	2.8	2.3	6.3	2.7	3.8	3.1	2.8	2.6	3.3	1.3
		2.9	2.4	6.4	2.8	4.0	3.3	2.9	2.8	3.4	1.3
		2.6	2.1	6.1	2.5	3.7	3.0	2.6	2.5	3.1	1.3
average per transect		4.0	3.3	9.0	3.8	5.5	4.5	4.0	3.8		

Population density estimation (deer/km²)

Studied area:	Casa Blanca, Oax.	
Date:	Dry season 2013	
Number of transects:	8	
Number of parcels:	80	
Total of pellet-groups:	108	
PELLET model		
Average density:	4.75	ind/km²
Estándar desviation (SD)	2.4	
Maximum:	13.9	
Minimum:	1.9	
Eberhart & Van Etten (1956) model		
Average density:	6.80	ind/km²
SD:	2.7	

column to select the entire column, hold down the shift key, and simultaneously click on the last transect column (AP in this example) to select all the remaining transects. In the top menu bar, click on Edit > Delete to remove these columns. Your new spreadsheet should now have only 10 transect columns but still have the green and yellow summary boxes. If you only have 5 transects, repeat the process to remove 5 more.

4. Enter your own data from the transect studies in Exercise 1. You may need to further edit the spreadsheet by removing excess cells.

5. As you enter new data, the spreadsheet automatically updates. When you are finished, you should see a new value for average density. Recall that this spreadsheet uses three estimates of defecation rate (12.2, 19.3, and 26.4). These values are for white-tailed deer. If you know defecation rates for your species, you can change them in the spreadsheet.

6. Write a scientific paper in the *Journal of Mammalogy* format describing your deer population and study site.

Exercise 3: Data Analysis Using Rdistance in RStudio

This tutorial assumes that you or your instructor have some familiarity with using Program R and with using RStudio, a graphical interface for R. Figure 11.3 shows RStudio's graphical interface.

This tutorial inputs data from two Excel files, fits a detection function, and estimates abundance of a hypothetical antelope population. Both Excel files can be downloaded from the Johns Hopkins University Press website for this manual. Data in the file called "antelope.xlsx" include 68 transects, each 500 meters long, where observers recorded the sighting distance and sighting angle to each antelope group detected. Open "antelope.xls" in Excel and take a look at the headings. They include a site ID, the number of antelope individuals in a detected group (groupsize), the distance from the group to the observer (sightdist or *r*), the sighting angle (sightangle or θ), and the perpendicular distance from the transect to the group (dist or *d*). The second Excel file you need is called "antelopeTrans.xlsx," and it

contains columns for site ID, transect length (500 meters), the observer, and the mean percentage cover at each transect.

The first step is to import the two Excel files into RStudio. To so this, open RStudio and click on the Import Dataset tab in the Data, Variables window (Figure 11.4). Click on From Excel, and navigate to the Excel file you want to use. The names of the two files should appear in that window when you are finished. You can also click on the file name in the Scripts window (upper left) to view the data. Alternatively, you can import the data sets using R commands as described below.

Next we move on to entering R commands in the Work Area window (lower left). The > symbol precedes the commands you enter. Commands shown below are in **Courier** font to make it clear what you type into RStudio at the > prompt. Be careful to enter commands exactly as shown. Enter the following commands (after each line hit return):

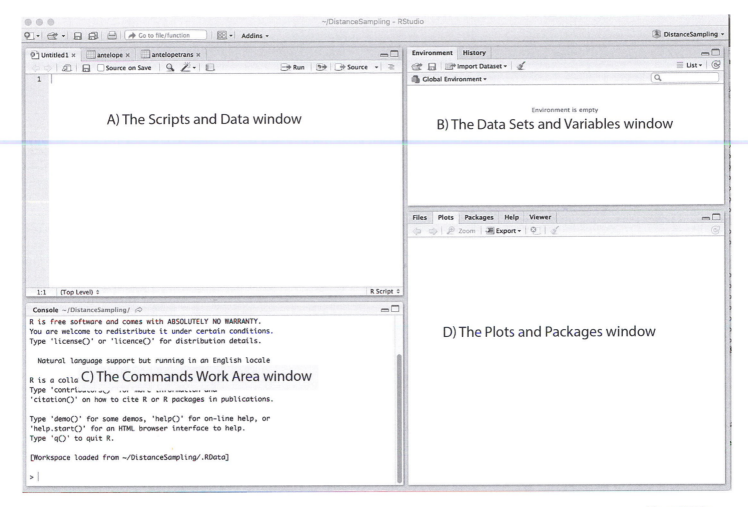

```
install.packages("Rdistance")
require(Rdistance)
```
After you enter both commands, the Rdistance package is loaded into your current session. When it is finished, you will see:
```
Loading required package: Rdistance
Rdistance (version 1.3.2)
```
The R program is now ready to do something with your data. Begin by loading your data using the following commands:
```
antelope=read.csv(file.choose())
data(antelope)
head(antelope)
```
You may get a warning message after the `data(antelope)` command, but continue on to enter the `head(antelope)` command.

Now we want to see the distribution of detection distances plotted as a histogram. Enter the following commands:
```
hist(antelope$dist, col="grey", main="",
xlab="Distance (m)")
```
This plots the detection distribution in the Plots window in the lower-right window in RStudio (Figure 11.5). The `col="grey"` tells R to plot grey histogram bars, and the `xlab=` is the *x*-axis label "Distance (m)."

Enter the command
```
summary(antelope$dist)
```
You should see values close to the ones below

Min.	1st Qu.	Median	Mean	3rd Qu.	Max.
0.00	14.62	31.00	40.47	59.08	207.00

Now we want to actually see the detection function. We will create it as a red line superimposed on our histogram. To do this, enter
```
dfunc<- F.dfunc.estim(antelope,
likelihood="halfnorm", w.hi=150)
plot(dfunc)
```
In this set of commands, dfunc is created using <– and it contains everything after that. You then plot the dfunc you just created. You should see a graph with the detection function in the Graph window (Figure 11.6). It also gives you the effective strip width (ESW).

The plot uses the half-normal likelihood as the detection function. We will try some other detection functions later in this exercise. Enter the command
```
dfunc
```
and you will see something like the following output:
```
Call: F.dfunc.estim(dist = antelope2,
likelihood = "halfnorm", w.hi = 150)
Coefficients:
   Sigma
49.87415
```

Figure 11.3 The RStudio interface showing the four windows mentioned in the text: (**A**) the Scripts and Data window, (**B**) the Data Sets and Variables window, (**C**) the Commands Work Area, and (**D**) the Plots and Packages window.

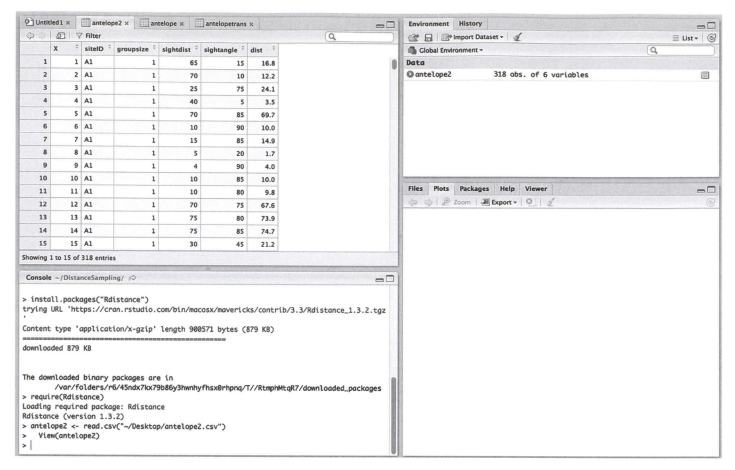

Figure 11.4 The RStudio interface showing that the data set "antelope2" has been imported into the Data Sets and Variables window and is available for viewing in the Scripts and Data window.

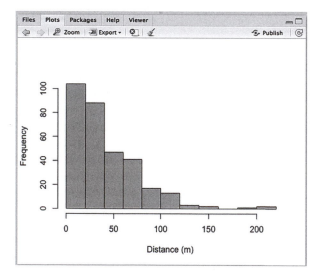

Figure 11.5 The histogram of detection distances shown in the Plots and Packages window. Note that you must click on the Plot tab to see the graph.

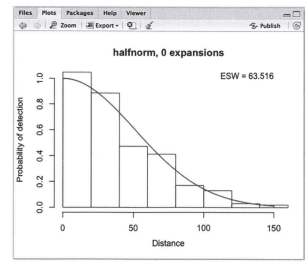

Figure 11.6 The histogram of detection distances with the half-normal likelihood detection function curve and the estimated strip width (ESW).

```
Convergence: Success
Function: HALFNORM
Strip: 0 to 150
Effective strip width: 62.34334
Scaling: g(0) = 1
Log likelihood: 1630.716
AIC: 3263.443
```

Our effective strip width (ESW) is 62.3 meters. We need ESW to calculate abundance. To calculate abundance/density, we also need the data in "antelopeTrans," so enter

```
antelopeTrans=read.csv(file.choose())
data(antelopeTrans)
head(antelopeTrans)
```

halfnorm, 0 expansions

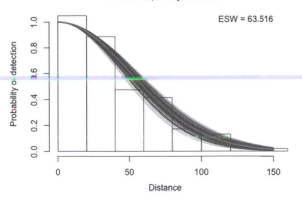

ESW = 63.516

Figure 11.7 The histogram of detection distances with several detection function curves and the new estimated strip width (ESW).

You will see an abbreviated table of the transect data. Now we will use the **F.adund.estim** command to estimate abundance. We also have to tell R the distance units used. If area=1 is used, then density is given in square meters. We want to use hectares (10,000 square meters), so enter the following command all on one line:

```
fit <- F.abund.estim(dfunc, detection.
data=antelope, transect.
data=antelopeTrans, area=10000, R=100,
ci=0.95, plot.bs=TRUE)
```

This command calls the **F.abund.estim** function and uses the data in antelope and antelopeTrans. It sets the area to square hectares with **area=10000** and the 95% confidence interval with **ci=0.95**. You should see your plot now has several detection functions on it (Figure 11.7).

Enter the command

fit

and hit return to get a summary of the information as shown below (your values may vary slightly):

```
Call: F.dfunc.estim(dist=antelope,
likelihood= "halfnorm", w.hi=150)
```

```
Coefficients:
  Sigma
49.87415
```

```
Convergence: Success
Function: HALFNORM
Strip: 0 to 150
```

```
Effective strip width: 62.34334
Scaling: g(0) =1
Log likelihood: 1630.716
AIC: 3263.443
```

```
Abundance estimate: 0.8751348;
95% CI=( 0.7665973 to 1.02088 )
```

This says there is an estimated 0.87 antelope per hectare, and we can be 95% confident that the true number is between 0.77 and 1.02 antelope per hectare.

If we want to use a different detection function instead of half-normal likelihood, we can use the **F.automated.CDA** command, which will attempt to fit a wide range of distance functions to the data. Are there better estimates than the half-normal likelihood? Enter the following command:

```
auto <- F.automated.CDA(detection.
data=antelope, transect.
data=antelopeTrans, w.hi=150, plot=FALSE,
area=10000, R=100, ci=0.95, plot.bs=TRUE)
```

You will see a large list of detection functions listed followed by

```
--------------- Final Automated CDS
Abundance Estimate--------------------
Call: F.dfunc.estim(dist=dist,
likelihood=fit.table$like[1], w.lo=w.lo,
w.hi=w.hi, expansions=fit.
table$expansions[1], series=fit.
table$series[1])
```

```
Coefficients:
    Beta          a1
0.02754839 -0.26542176
```

```
Convergence: Success
Function: NEGEXP with 1 expansion(s) of
COSINE series
Strip: 0 to 150
Effective strip width: 44.72749
Scaling: g(0) =1
Log likelihood: 1628.032
AIC: 3260.098
```

```
Abundance estimate: 1.219805 ;
95% CI=( 1.030589 to 1.675932 )
```

The "best" detection function is NEGEXP or negative exponential likelihood (do not worry about the expansions for now). The best estimate of abundance is now 1.2 antelope per hectare with 95% confidence intervals of 1.03 and 1.67.

12 Camera Trapping

Time Required

- The initial data collection phase of this exercise can be run over several weeks or months. Data analysis can be completed in 3–4 hours.

Learning Objectives

- Learn how to use cameras as remote sensors in wildlife studies.
- Understand how to design and implement a camera trapping study.
- Gain experience placing cameras in the field.
- Learn how to collect and analyze camera trap image data.
- Use RStudio software to analyze animal populations from camera trap data.

Equipment Required

- Camera traps (number depends on objectives and size of study area)
- Extra batteries for cameras
- SD data storage cards (minimum of 1 per camera)
- SD card picture viewer (optional)
- GPS device

Background

Conservation biologists increasingly rely on rapid assessments of species richness and abundance, especially in regions threatened with habitat loss. These quick-and-dirty surveys are crucial for assigning conservation priorities. In the past these surveys have relied on track counts, short periods of livetrapping, or direct observation of animals along line transects. Although each method has its advantages, they all require considerable time and effort and do not work well for cryptic or wide-ranging species. Remotely triggered cameras (e.g., camera traps) have been used with great success in recent years to survey large tracts of remote areas (Ancrenaz et al., 2012; Jansen et al., 2014; Shannon et al., 2014). Networks of motion-/heat-sensitive cameras are an efficient and cost-effective way to collect data for species inventories and are especially important for recording

the presence of cryptic animals (Balme et al., 2009; Karanth and Nichols, 1998).

Camera trap networks

- are non-invasive;
- are relatively simple to deploy;
- operate without supervision for weeks or months;
- provide permanent records of species, date, and geo-referenced locations;
- can record animal behavior, if video camera traps are used; and
- are cost-effective over the long term.

In addition to documenting presence of a species at a site (Figure 12.1), a properly designed camera trapping study, using carefully chosen, multi-year, geo-referenced locations can also document species distributions, local density (Rowcliffe et al., 2008), habitat preferences, and other population demographic parameters (Karanth et al., 2006). Therefore, camera trapping is rapidly becoming a critical tool for wildlife conservationists (Karanth, 1995; Kays and Slauson, 2008).

The most successful camera trap studies are those that are well planned in advance. Researchers need to

Figure 12.1 Photo of a white-tailed deer taken with a camera trap in New York State. Photo courtesy of J. Ryan

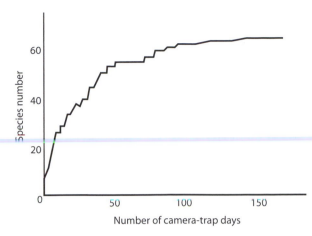

Figure 12.2 An example of a species accumulation curve. Notice that after approximately 75 days few additional species are added to the survey.

know what information they will need to answer their research or conservation question. In addition, it is vital to have knowledge of the geographic area and terrain to be sampled. Researchers should plan approximate camera positions and travel routes in advance. It helps to know the approximate home range size and basic ecology of the target species. Obviously, knowing how to properly operate the camera trap units and a GPS device is also vital. In most cases, a pilot study is required to become familiar with equipment, to learn the logistics of moving through the terrain to check traps and replace batteries, and to get an estimate of capture success (number of photos or videos per unit of time).

A research project that aims to document which species are present at a particular site (i.e., to produce an inventory) should attempt to maximize the chance of acquiring clear images of each species. This means deploying cameras in all habitat types. Typically, cameras would be placed along game trails, near water sources, or at bait stations. In addition, cameras using white Xenon flashes are best suited to species identification for many medium-size to large mammals. In this type of study, cameras are deployed long enough to allow species accumulation curves to level off (Figure 12.2).

In contrast, a study that aims to determine the population density of a target species would use a different approach. Density estimates usually require recaptures, which means the goal is to maximize the opportunity to take repeated photos of the same identifiable individual. This approach works well with spotted cats, whose unique patterns of spots allow each individual to be recognized. In this study design, researchers may place two or more cameras at each site to ensure photos from different angles. Cameras would be placed over an area large enough to sample a reasonable proportion of the population (i.e., 10-30 individuals of a target species).

If the plan is to use occupancy models to assess spatial distribution, then researchers may want to deploy

cameras using a grid design that does not attempt to locate cameras in particular habitats or sites. Protocols for optimizing camera-trap placement to meet the study's objectives are numerous (Ferreras et al., 2017; Hamel et al., 2013; O'Brien & Kinnaird, 2011; Xingfeng et al., 2014), and the key point is to carefully consider your study objectives before deploying camera traps.

Camera Selection

Choosing among the dozens of models and types of digital camera traps on the market can be daunting. In the end, the choice will be dictated by the requirements of the study, the budget, and the preference of the researcher. There are, however, some basic considerations. Digital cameras have all but replaced film cameras in recent years. Digital camera traps offer a number of advantages, such as longer battery life, infrared flash to capture nighttime images, and the capacity to store hundreds of images on tiny memory cards with no film development required.

Digital camera traps can be triggered to capture an image using either active infrared (AIR) or passive infrared (PIR) triggers. Cameras using AIR capture an image when an animal crosses a narrow infrared beam; cameras using PIR rely on the difference in temperature between an animal and its ambient surroundings. AIR camera traps are very effective at capturing target animals, but windblown leaves, flying birds, and rainstorms can also trigger the cameras. PIR traps can have false captures as well.

Camera Features to Consider When Choosing Camera Traps

Cost is always a consideration, but as with most things in life, you get what you pay for. In 2017, digital camera traps ranged from about $70 to more than $1,100. The highest-priced cameras come with many important features, including fast trigger speeds and high megapixel ranges (more megapixels means larger, sharper images). Before you buy the cheapest or most expensive model, carefully evaluate your needs. For example, you do not need a 5+ megapixel image to identify an animal, but faster trigger speeds and burst photo modes may be critical for some studies. With typical camera trap surveys using 20-100 camera traps, cameras can quickly become a substantial budget item.

Trigger speed is the time from when the infrared trigger first detects motion or temperature until the camera captures a photo. Rapid trigger speed is probably the most important consideration for large mammals. If it takes the camera 3 seconds to power up out of sleep mode and take a picture, the animal may already be gone (or you end up with a photo of its rump). Trigger speeds vary widely, with cheaper units having speeds as long as 6 seconds (useless for scientific studies). Trigger speeds of 1/2 second or less are best for wildlife surveys. If you must choose a camera with a slower trigger speed, make

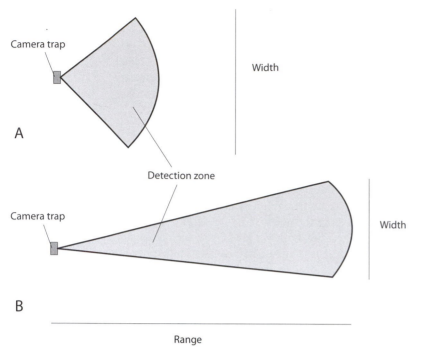

Figure 12.3 Comparison of detection zones from camera traps.

sure it has a wide detection zone so that the camera has a higher probability of taking a photo when the animal is still within the zone. Slower trigger speeds may still be useful at baited stations.

A camera trap's **detection zone** is the area within which movement is detected. Some cameras have narrow and long or short and wide detection zones (Figure 12.3). For most field studies, a wide detection zone matched to the camera's lens field of view is best.

Recovery time is the minimum time between two consecutive photos (in a non-burst mode). Recovery times vary from 1/2 second to 60 seconds. Researchers should consider whether one photo every 60 seconds is adequate for their study design. For example, consider a scenario where a mother tiger triggers the camera and is followed in the next few seconds by two cubs, who pass before the camera recovers. Such a scenario would result in the loss of important information for tiger conservationists. Recovery times of 1 second or less are acceptable in most cases. Rapid-recovery cameras often capture multiple images of each animal, which may be important if the animal carries visible tags or has a unique coat pattern for identification. (Remember, larger-megapixel pictures require additional time to write to memory.)

Flash range is the maximum distance the flash projects to illuminate an animal with sufficient intensity to identify the animal. Manufacturers often report optimal values (up to 80 feet in some cases), but those are usually taken in ideal situations (e.g., open fields or in moonlight). The same camera located in a deciduous forest or on a cloudy night would lose as much as 50% of its advertised flash range. Infrared flashes are also available and less invasive to most mammals.

Manufacturers give **battery life** as the number of days a camera is able to operate on a single set of batteries. Many variables affect battery life: the number of pictures taken per day, ambient temperature, camera settings, and many other factors. Consider using rechargeable NiMH, NiCd, or Li-ion batteries for longer battery life. Always test your batteries before placing them in the unit; many over-the-counter battery packages contain at least one dead battery.

Other Factors to Consider in Choosing the Best Camera

Some camera traps have more features and require more technical expertise for proper field use. Consider who will be deploying the cameras and the level of training required. Some camera traps are weatherproof units that can be submerged for short periods, but many units are only modestly water resistant. Consider the environment and weather conditions carefully before purchasing camera traps.

Recent advances in video technology have reduced the size and cost of video devices to the point where many researchers are opting to use video camera traps. Many of the digital camera traps discussed previously also have the capacity to take short video clips and store those to memory. These may be pseudo-video if the number of frames per second is less than the standard 30 fps for video. In addition, the resolution of the video footage is often only 640 × 480 pixels per frame. For researchers wanting behavioral information, video capability may be a necessity. As always, there are drawbacks, including shorter battery life, fewer events recorded to memory, longer trigger times, and greater expense.

Survey Design

There are a number of criteria to consider in designing a camera-trapping study. Foremost among them is meeting the study objectives. Often there are trade-offs involving the number of cameras available, the duration of the study, the coverage area, and statistical or other data analysis requirements. Deployment duration (days in use) and site coverage (size of the survey area) are primary factors in designing effective studies. The longer the deployment, the greater the probability of detecting a given species, but longer deployment at one site also means that fewer sites can be surveyed in a given period of time (e.g., one season).

Camera traps have been used to address a variety of research issues, such as the estimated density of target species, the relative abundance of target species, the relative abundance of mammal communities, species diversity, and activity patterns, as well as to document the use of a specific habitat feature (e.g., burrow entrances or underpasses). Cameras can then be used to compare any of these metrics between different study areas (i.e., treatments). In some circumstances, bait stations near camera traps may be the preferred method of documenting the presence of a specific species.

The next decision is how long to deploy cameras (Xingfeng et al., 2014). This will depend on the remoteness of the site, logistics, and other considerations. If the goal is to estimate density using a closed population model, then cameras should be in the field just long enough to capture sufficient data without violating the model's assumptions. For example, mark-recapture models may require a closed population (i.e., no individuals can be added or removed by birth, death, or dispersal during the study). Long-duration studies may violate this assumption. A second assumption of these models is that every individual has a nonzero probability of being captured (photographed). This is the main reason for spacing cameras in arrays with inter-camera intervals less than the minimum home range area for the target species. This does not mean that every individual must be photographed, only that each individual has the same chance of being photographed.

Exercise 1: A Camera-Trap Field Study

Design a camera-trapping protocol for a site selected by your instructor. Use the information provided here to determine camera placement and overall study design. Collect SD memory cards (digital photos), and replace batteries as needed on a regular schedule. Organize the photos into a database or spreadsheet file for analysis.

Before you go to the field, study a topographic map or satellite photos of the survey area. Locate and mark all trails, roads, waterholes, streams, and other features where there is a high likelihood of photographing your target species. Begin locating potential camera stations on the map by penciling in locations and spacing them appropriately.

Now look for gaps between cameras greater than your study design allows and move camera stations around to fill those gaps (e.g., you may need to move cameras owing to lakes or other non-habitat within the study site). Once you have a potential map of your camera stations pinpointed on the satellite map, use Google Earth to record the exact coordinates (latitude and longitude) of each station. These "predetermined" coordinates will serve as a guide to finding the sites in the field using a hand-held GPS unit.

Remember:

- Cameras must be monitored, so consider the logistical implications of your survey map.
- Camera placement need not be random; the final field position may differ slightly from those on your survey map.

Upon arrival at a predetermined camera station at your field site, look over the area carefully. Each location offers unique challenges for positioning the camera traps. Find the best possible location as close as possible to the predetermined coordinates. Look for signs (e.g., tracks, scat) or game trails. Many mammals regularly use trails, dirt roads, riverbanks, waterholes, and game paths. Choose sites where animals are likely to pass within range of the camera. Avoid open areas where the animal's path is difficult to predict. Keep in mind that humans may also follow these paths, and human presence may prevent animals from using these paths regularly. Humans may also remove or damage cameras. Spending even a few hours to find just the right spot is well worth the effort, since each camera will be in the field for weeks or months.

Once you have found a suitable site for the camera station, take the following steps:

1. Take a GPS reading and record the location of the station.

2. Select one or two suitable trees to which to attach the units. Trees should be relatively straight and thin enough to tie the traps in place (but not so thin as to move with the wind). Trees should be two meters from the most likely travel path.

3. Make sure the ground is reasonably level and there are no obstructions in the field of view of the camera.

4. Position cameras should be slightly down trail (rather than exactly perpendicular).

5. If using PIR traps, avoid sites where cameras will be in direct sunlight during sunrise or sunset.

6. Consider the height of the animals, and position the traps about 50–100 centimeters from the ground.

7. Make sure that the date and time stamp is set correctly (photos are essentially useless without a date and time).

8. Load the units with new batteries that you have tested.

9. Set the programmable features to the desired settings (e.g., image size, trigger delay).

10. Secure the camera traps to the tree using wire or chain. Attach a lock to the camera to prevent theft. For baited stations, aim the camera at the bait station and secure the bait so that the animal cannot easily remove it.

11. Attach a label that indicates the purpose of the study and provides contact information.

12. Test cameras by walking in front of them (most camera traps have an indicator light that lights up when something is detected).

13. Describe the station in your field notes.

Data Analysis

Photos are collected from each camera during routine monitoring of the study area. Photos can be collected from digital camera traps in several ways:

- Remove the SD memory card, and replace it with a new one. Label these used SD cards, return them to the lab, and download them onto a computer using a standard card reader.
- Bring a computer into the field, and use a USB cable attached to the camera trap to download the photos.
- Use a portable SD card viewer to visually scan each SD card at the station and transfer the images to a second SD card.

Good record management is critical to a successful study. Each SD card should be labeled with its station and camera number. Images from each camera should be entered into a database, keeping such information as camera number, station number, date, time, and species as separate fields in the database.

Camera-trap data consist of date- and time-stamped images of a particular species at a geo-referenced location. Ideally, you would like to be able to identify individuals, but this is not always possible. For spotted and striped cats (e.g., jaguars, ocelots, leopards, tigers), patterns of spots or stripes on the body can be used to iden-

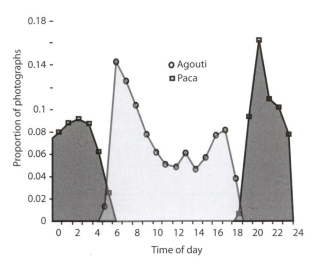

Figure 12.5 Activity patterns of two tropical rodents on Barro Colorado Island, Panama. Notice the diurnal and nocturnal patterns of each species. *Data from Kays et al. (2011)*

tify individuals from photos. For more uniformly colored species (e.g., ungulates, coyotes, mountain lions, bears), it is generally not possible to identify individuals from photographic data. Nevertheless, camera-trap data can be used to estimate density even without individual recognition (Rowcliffe et al., 2008). These estimates, however, are complex and beyond the scope of this discussion.

One of the most basic measures is detection frequency (Figure 12.4), often defined as one photograph of a species per camera trap per day (24 hours). The date and time stamps on each photo also provide information about activity patterns (Figure 12.5).

Another relatively simple measure is detection rate, the total number of photos of a species divided by the total time a camera was running (excludes periods of camera malfunction). Summary information, including effort or trap-nights and species richness, should also be tabulated for each study. A relative abundance index (RAI) can also be calculated from camera traps. The RAI for each species is calculated by summing all detections for each species for all camera traps over all days, multiplied by 100, and divided by the total number of camera-trap days. For example, if 10 cameras are used for 100 days, then $10 \times 100 = 1,000$ camera-trap days. If one black bear was photographed four times on day 1, one time on day 4, and one time on day 5, then the RAI for the bear is calculated as

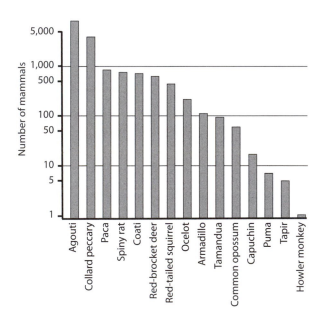

Figure 12.4 Frequency of detection for 15 tropical mammals on Barro Colorado Island, Panama. *Adapted from Kays et al. (2011)*

$$RAI = \frac{(1+1+1)(100)}{10 \times 100} = \frac{300}{1,000} = 0.3$$

Exercise 2: Data Analysis from Camera-Trap Studies

Using the data you collected from your field site in Exercise 1, calculate the species richness, detection frequency of each species, detection rate, RAI, and activity patterns of species. Display your data graphically. Write a "Materials and Methods" and a "Results" section for a research paper that you might submit using this study design (follow the format for the *Journal of Mammalogy*; download the "information for contributors" document at https://academic.oup.com/jmammal/pages/General_Instructions).

If the target species carry unique coat markings (e.g., spot or stripe patterns), additional information can be gleaned from camera-trap studies using mark-recapture methods (Karanth and Nichols, 1998). With large data sets, it is more practical to use computer software programs to help analyze camera-trap data.

Exercise 3: Using camtrapR to Analyze Camera-Trap Data

You will use an R package developed for analyzing camera-trap data called camtrapR (Niedballa et al., 2016). Begin by launching RStudio and installing the camtrapR package by typing the following commands at the > prompt in the Console window (lower left) in RStudio.

```
install.packages("camtrapR")
library(camtrapR)
require(camtrapR)
```

In RStudio, you enter commands after the > prompt in the Console window or as a script in the script window. The > prompt is not shown, just the commands that follow it. In a script, you will often see the # symbol followed by comments that remind you what you are doing. Text after the # prompt is not a command and is not executed by the program. Thus, you might type

```
# load sample camera trap station table
data("camtraps")
# load sample record table
data("recordTableSample")
```

These lines represent two comments, each followed by one command.

Begin by generating a map of the species richness at each station. We will use the detectionMaps function. Enter the following as one long record without hitting return:

```
Mapstest1 <- detectionMaps(CTtable =
camtraps,
    recordTable  = recordTableSample,
    Xcol         = "utm _ x",
    Ycol         = "utm _ y",
    stationCol   = "Station",
    speciesCol   = "Species",
    printLabels  = TRUE,
    richnessPlot = TRUE,
    speciesPlots = FALSE,
    addLegend    = TRUE)
```

This is a long command. It creates a new dataframe called Maptest1 and plots the position of the camera-trap stations using utm coordinates on the *x*-axis (Xcol) and *y*-axis (Ycol). Each station is labeled with a dot that

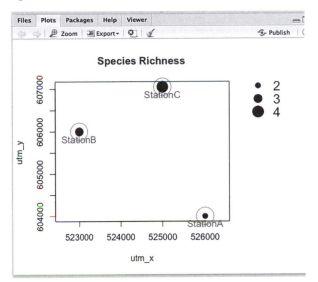

Figure 12.6 A species-richness plot from the camtrapR package for a small data set with three trap stations and five species.

varies in size depending on how many species were detected at that station (Figure 12.6).

You can replot the graph using one species as a time. In this case, you are plotting the number of detections for that single species at each station. There are five species in this small data set. Begin by plotting the data for the leopard cat (*Prionailurus bengalensis*, PBE):

```
Sample _ PBE <- recordTableSample[record
TableSample$Species == "PBE",]

Mapstest2 <- detectionMaps
(CTtable = camtraps,
    recordTable   = Sample _ PBE,
    Xcol          = "utm _ x",
    Ycol          = "utm _ y",
    stationCol    = "Station",
    speciesCol    = "Species",
    speciesToShow = "PBE",
    printLabels   = TRUE,
    richnessPlot  = FALSE,
    speciesPlots  = TRUE,
    addLegend     = TRUE)
```

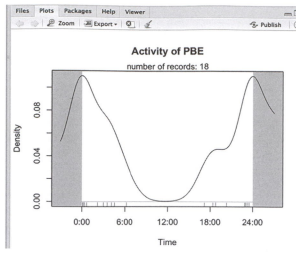

Figure 12.7 The locations of trap stations where the leopard cat (PBE) was detected using the sample data from the camtrapR package.

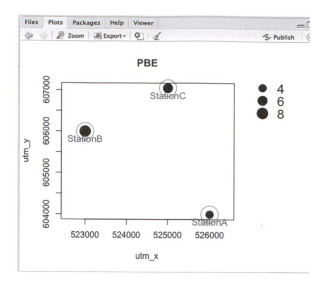

Figure 12.8 The activity pattern of the leopard cat (PBE) from camera-trap data in camtrapR.

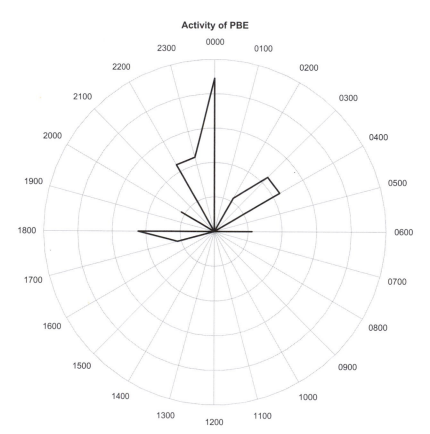

Figure 12.9 A radial plot of the same data shown in Figure 12.8.

The plot should look like Figure 12.7.

Suppose you are more interested in the activity patterns of one or two species in the study area. You can use the activity functions to do this. Begin with one species, the leopard cat (PBE).

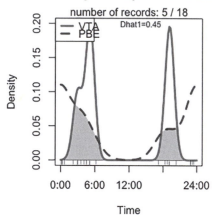

Figure 12.10 Activity patterns of the leopard cat (*dotted line*) and the Malay civet (*solid line*) plotted together in camtrapR.

```
PBEactivity <- "PBE"
activityDensity(recordTable = record
TableSample, species = PBEactivity)
```
This plots the activity patterns for the leopard cat (PBE) over time (Figure 12.8)

If you want it plotted as a radial graph, use (again all entered on one line).

```
activityRadial(recordTable = record
TableSample,
   species           = PBEactivity,
   allSpecies        = FALSE,
   speciesCol        = "Species",
   recordDateTimeCol = "DateTime
     Original",
   plotR             = TRUE,
   writePNG          = FALSE,
   lwd               = 3,
   rp.type           = "p")
```

This generates the plot in Figure 12.9.

To plot two species at the same time, type

```
civet <- "VTA"
leopard _ cat <- "PBE"
activityOverlap
(recordTable = record TableSample,
    speciesA   = civet,
    speciesB   = leopard _ cat,
    writePNG   = FALSE,
    plotR      = TRUE,
    linecol    = c("red", "blue"),
    linewidth  = c(3,3),
    add.rug    = TRUE)
```

Figure 12.10 shows the new two-species activity plot. Lastly, we'll look at some summary data from the data set. Enter the commands

```
data(camtraps)
data(recordTableSample)
reportTest <- surveyReport
(recordTable = recordTableSample,
    CTtable    = camtraps,
    speciesCol = "Species",
    stationCol = "Station",
    setupCol   = "Setup _ date",
```

```
    retrievalCol           = "Retrieval _ date",
    CTDateFormat           = "%d/%m/%y",
    recordDateTimeCol      = "DateTime
        Original",
    recordDateTimeFormat = "%Y-%m-%d

%H:%M:%S")
class(reportTest)
length(reportTest)


reportTest[[3]]
reportTest[[4]]
```

It prints the number of detection events and number of stations by species. As you can see, the leopard cat (PBE) was detected 18 times at three stations. The reportTest 4 shows the number of times each species was detected at each station. The leopard cat was detected at all three stations (A–C).

The camtrapR package can do much more, including managing photos and exporting data in GIS formats. Feel free to explore the package by referring to the reference manual, which can be found at https://cran.r-project.org /web/packages/camtrapR/index.html.

13   Radio Tracking

Time Required
- One 3- to 4-hour lab period for collecting the radio tracking data and one 3- to 4-hour lab period to analyze the data.

Learning Objectives
- Understand the principles of remotely tracking mammals.
- Learn how to capture, immobilize, and collar small mammals for a radio tracking study.
- Understand how to acquire triangulation data.
- Understand how to analyze radio tracking data.
- Understand how radio tracking is used to study mammals in the field.

Equipment Required
- VHF radio collars (minimum of one)
- Receiver tuned to collar frequency
- Yagi handheld antenna tuned to collar frequency
- Sighting compass
- Two-way radios for communication between tracking stations
- Livetraps (optional)
- Immobilizing drugs and equipment (optional)
- Field notebook (optional)
- Permits and Institutional Animal Care and Use Committee (IACUC) approval
- Tracing paper, compass circles, and straightedges
- Computer

Background

Three main types of telemetry are used in wildlife tracking studies today: (1) radio tracking using VHF (very high frequency) signals; (2) satellite tracking using AR-GOS or other satellite systems; and (3) Global Positioning Systems (GPS) data loggers or transmitters. VHF wildlife tracking, in use since the early 1960s, is the focus of this lab (Kenward, 1987; GPS tracking will be discussed in Chapter 14). Wildlife radio telemetry typically involves the transmission of radio signals sent from an animal carrying a radio transmitter to a researcher hold-

Figure 13.1 A recently radio-collared wolf in Yellowstone National Park. *Photo courtesy of William Campbell, US Fish and Wildlife Service*

ing an antenna and radio receiver tuned to the frequency of the transmitter (Figure 13.1).

Radio telemetry can address many research questions. It can provide information on an animal's location, movement patterns, home range, migration patterns, habitat preferences, den or hibernaculum sites, and in certain cases physiological parameters such as body temperature and dive depths (for marine mammals). Nevertheless, before embarking on a telemetry study, researchers should ask themselves:

- Is radio tracking the best method available?
- How many animals must be tracked to acquire sufficient data?
- Can the species be readily captured (and recaptured)?
- Can the study species carry the transmitter without altering its behavior or survival?
- What size radio transmitter will be needed for the duration of the study?
- How often will subjects be tracked?
- How many people will be required to track the subjects of the study?

- Is the budget sufficient to cover the costs of transmitters, receivers, and personnel?

Among the most important of these questions are those involving sample size—both the number of animals to be tracked and the number of positions for each animal tracked. For example, 100 tracking positions could come from 1 animal, or from 10 animals each with 10 positions. Before conducting the study, researchers must consider the sample size to ensure that appropriate statistical tests can be used on the data collected. In general, it is the number of animals tracked that is most important (to avoid pseudo-replication), but compromises must be made between tracking many individuals and collecting a sufficient number of positions for each individual. The objectives of the study, then, must drive decisions about sample size. For example, studies on home range size and overlap may need to sample adequate numbers of both males and females.

Alternatively, studies that seek to understand migration patterns or seasonal use of habitats may need to track individuals over many months. It is important to remember that statistical tests often require that radio positions be a random sample; this generally requires acquiring positions at random times throughout the study period or, more commonly, sampling at regular intervals. In certain cases, radio tracking is used just to locate an individual or its group in order to conduct behavioral observations. In such cases, only one individual in the group needs to be tracked and the number of exact positions is less important (excellent discussions of telemetry can be found in Kenward, 1987).

Types of Radio Telemetry Studies

One of the most common uses of radio telemetry is to understand habitat usage. The proportion of radio positions in each habitat relative to the proportion of habitat available can indicate habitat preferences (or avoidance). Care should be taken to ensure that both sexes are radio tagged, as habitat use often differs between the sexes. It is important to collect radio positions during the full 24-hour day if the study involves species with nocturnal or crepuscular habitats, or habitats that differ between day and night. Habitat preferences may also vary dramatically between years. Ideally, a fixed schedule of sampling should be employed. Each radio position must also be accompanied by a habitat classification.

Habitat can be classified in the field during the radio tracking process, or later, when the researcher revisits each location. Alternatively, the positions can be plotted on a detailed habitat map of the region (if one exists). One important source of error in such studies concerns the size of the habitat patches and the size of the telemetry errors. Each radio telemetry position is the result of triangulation and has an error polygon associated with it (Figure 13.2). If the habitat patches are smaller than the

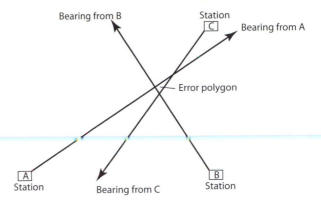

Figure 13.2 Error polygons associated with triangulation of radio telemetry positions vary in size and must not be larger than habitat patch size.

telemetry error polygon, then the proper assignment of habitat is difficult (Nams, 1989).

A common use for radio telemetry is to home in on a group of animals in which one individual is outfitted with a radio collar. In this case, the main objective is to reduce the time spent locating the group. Radio tracking is an excellent way to locate dens, roosts, or feeding areas. Biologists, however, should take extra care when attaching transmitters to breeding individuals to ensure no disruption to the animal's survival and/or reproductive success. For example, pregnant females may be especially susceptible to the stresses associated with capture and transmitter attachment. In addition, following tagged animals to den sites may expose the litter to increased predation or observer disturbance.

Another common use for radio-tracking data is to determine home range size for several representative animals in a population. If there is little variation in home range size among individuals of different age and sex categories, then the data may be pooled from these groups. However, males often have significantly larger home range size than females. In such cases, it would not be appropriate to pool data from both sexes. Kenward (1987) suggested that approximately 30 relocation positions per individual is an adequate sample size for home range determination. In general, it is best to determine the appropriate sample size for a study of home range size or other kinds of studies by conducting a pilot study or constructing an asymptotic plot of home range size versus number of positions. In practice, an animal's home range size and shape are determined by constructing a minimum convex polygon from the positions, or by using software for fitting polygons or kernel estimates (Seaman and Powell, 1996; Seaman et al., 1998). Similarly, daily movements (distance and velocity) can be summarized from the distances between consecutive radio positions.

In theory, radio telemetry data should help determine the cause of mortality, if the collared animal can be located soon after death. Survival rates can also be calculated from the number of transmitter-days (Pollock et al., 1989). In practice, it is often difficult to determine the cause of mortality or to distinguish mortality from tag loss or tag failure.

Radio Telemetry Equipment

Conventional radio transmitters consist of a transmitter unit, a battery (or solar cell), and an antenna. The specific components chosen for a study will depend on the study design and the size of the study species. The transmitters (also called "tags") are typically purchased as complete units packaged in acrylic or epoxy resin to protect them from weather or the teeth of the study animal. Radio transmitters come as one-stage or two-stage units. One-stage transmitters consist of an oscillator and a power source. They are simple and lighter in weight and have longer battery life for a given tag weight. Two-stage transmitters have both an oscillator and an amplifier. These tags require more power to operate (>2.4 volts) and are therefore larger, weigh more, have shorter battery life for a given tag weight, and are more expensive. On the other hand, two-stage transmitters have greater range (usually almost double). Animals that are small or have relatively small home ranges are good subjects for one-stage transmitters. The transmitter (with power source) must not weigh more than 3–5% of the subject animal's body weight.

Transmitters are powered by either a lithium or silver battery or by a solar cell. Solar-powered transmitters have obvious limitations for studies involving nocturnal, fossorial, or aquatic species. Lithium batteries are the most common source of power. Battery life and signal range are inversely related. Range is generally reported as line-of-sight range and assumes open, level ground with no obstructions. In practice, range is reduced by weather conditions, dense foliage (e.g., tropical forests), uneven terrain, and reflection off water bodies. In general, larger batteries yield increased range and increased battery life, but at the expense of greater weight.

Transmitters can be turned on and off by magnetic switches. They are shipped with a magnet taped to the outside of the unit (in the "off" configuration) and activated when the magnet is removed in the field. For transmitters with small batteries and short battery life, a magnetic switch may conserve battery life, but they are more expensive and add to the size and weight of the transmitter. Alternatively, transmitters can be shipped with an unfused connection; the biologist solders the connection closed to activate the unit and covers the connection with epoxy in the field.

Each transmitter has a small internal or external antenna attached to the unit. Whip antennas are stainless steel wires coated in Teflon that extend outside the transmitter unit. Whip antennas are light, strong, and omni-directional, and produce a relatively uniform signal. In contrast, loop antennas produce their maximum signal when they are tuned to the circumference of the animal's neck (the antenna is generally incorporated into the collar material). Loop antennas produce weaker signals but are less subject to breakage or chewing.

Radio telemetry may also provide information on activity or certain physiological parameters. Activity sensors use pulse interval modulation (PIM) to represent the animal's activity. For example, real-time PIM sensors use a mercury tip-switch to send radio signals of one pulse rate when the mercury in the switch is in one position and a different pulse rate when the mercury bead moves to the opposite position in the switch (e.g., when the animal moves, tipping the switch). Pulse rates that vary tell the researcher that the animal is active. Time-delay activity sensors also use a tip switch, but in this case the switch has a counter. The transmitter's pulse rate changes only if the switch is not triggered within a specified period of time. Such sensors are commonly used to detect mortality or tag loss. Temperature sensors incorporated into transmitters are used to monitor the animal's body temperature. Body temperature transmitters are usually placed subcutaneously or intra-abdominally. As body temperature varies, the pulse rate of the transmitter also varies. Thus, it is critical to carefully calibrate the unit ahead of time over a known temperature range. Light transmitters are used to detect the amount of light reaching the transmitter and are typically used to determine activity periods for burrowing species.

Transmitters may be attached to wildlife in several ways depending on the animal's body size, shape, lifestyle, and the needs of the researcher. (Before you begin, be sure you have an approved IACUC protocol and proper state and/or federal permits, and consult a veterinarian about immobilization procedures.) Collar attachment is one of the most popular as it provides a relatively comfortable and durable attachment to the animal (Figure 13.1). Collars should be made of materials such as butyl belts, urethane belts, or nylon webbing materials. In certain cases (such as for small mammals), metal ball-chains or cable ties are also used. Properly fitting collars to animals takes some experience. Collars are typically placed so that the transmitter hangs below the neck. They should fit snugly but not so tight as to be uncomfortable (e.g., they should not hamper breathing). Manufacturers are good sources of information for data on neck circumference in various mammal species. Some manufacturers provide breakaway collars that eventually break and allow the collar to fall off. These collars are useful if the researcher believes that recapturing the animal may be difficult.

Implantable transmitters are placed inside the animal's body (i.e., subcutaneously or intra-abdominally). They are typically used for animals with necks thicker than the width of the skull (e.g., small mustelids, mongoose), for burrowing mammals where a collar may rub against the tunnel walls, or for young mammals that are expected to grow substantially over the course of the study. Implantable transmitters are also used when the study design requires data on body temperature. While implantable transmitters are more comfortable (if implanted properly) for the animal, their range is limited, and the subject animal needs to be held in captivity until it recovers from the implantation surgery.

Backpack transmitters are secured to the animal's back by a harness made of plastic-coated wire, nylon webbing, or other material. The style of backpack harness depends on the study species. Backpacks may snag on vegetation and can cause chafing; they are used when other methods are not appropriate.

In some cases, transmitter assemblies are glued to the animal with cyanoacrylate glue, surgical adhesive, or other substances. Transmitters are typically glued directly to the skin (the fur over the attachment area is removed) or directly to the fur (e.g., bats, voles). Glued transmitters are positioned so that the transmitter is on the dorsal surface (the back) with the whip antenna extending toward the tail. These tags generally detach themselves over time as the fur molts or during grooming; this attachment method is generally suitable for shorter-duration studies. There are also a variety of specialized attachment systems, including ear-tag transmitters (e.g., for large mammals), and new techniques are in development.

The following are some general rules of thumb:

- Use the smallest tag possible to accomplish the goals of the study. Use tags weighing less than 3–5% of the animal's body weight.
- Test the attachment method on captive animals before the study.
- Tag at least two animals in large social groups in case one tag fails prematurely.
- Avoid placing tags over markings that the animal uses for intraspecific communication.
- Test each tag before attaching it to the animal to ensure it is working properly.
- Follow all guidelines for properly immobilizing animals (or have a veterinarian on hand).
- Allow the animal several days to adjust to wearing the collar before collecting data.
- Treat each animal with care during the collaring and tracking process.

Radio tagged animals carry a transmitter that sends out a pulsed signal of a given frequency. The researcher carries a receiver unit that picks up the signal via a hand-held antenna and amplifies the signal so the researcher can hear it (using earphones). Receivers come in many sizes, weights, and prices; the choice will depend on the study's objectives and budget (Figure 13.3). Receivers are powered by batteries (replaceable or rechargeable) and also usually come with an adapter for connecting to a vehicle.

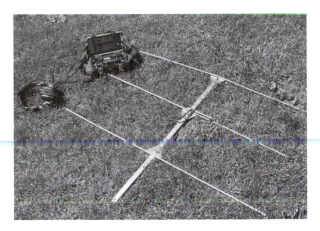

Figure 13.3 Radio tracking receiver, headphones, and a 3-element Yagi antenna. *Photo courtesy of J. Ryan*

Receivers can detect a range of frequencies and can thus be tuned to the transmitter frequency range. Each transmitter frequency is entered into the receiver before the area is scanned for a signal (some models have scanners that automatically switch between a number of different frequencies). Like all electronic circuitry, receivers can be damaged by humidity, water, or excess static electricity (e.g., from vehicle seats and clothing).

Handheld antennas are attached to the receiver with coaxial cable. The most common antennas are the Yagi or H antennas. These directional antennas concentrate their energy toward the front of the antenna (some energy also goes to the sides and behind). The beam width of the antenna is determined by the orientation (vertical or horizontal) and the number of elements in the antenna. For example, a three-element Yagi antenna will have a narrower beam width if held in a horizontal position (approximately 60 degrees) rather than a vertical position. Likewise, this same three-element antenna will have a narrower beam width than a two-element antenna held in the same position (roughly 100 degrees). Loop antennas are used for close-range tracking (e.g., 1 kilometer or less). Larger antennas (up to 14 elements) are sometimes mounted on vehicles, boats, or aircraft or used as part of a permanent ground station.

Locating Animals

Two methods are typically employed to locate tagged animals. The first, called homing, involves using the receiver and antenna to find the direction in which the signal sounds the loudest. A compass bearing is taken along the line at which the antenna is pointing when the signal is loudest. The researcher follows that line, taking additional readings and adjusting the travel bearing until the animal can be seen.

Alternatively, the process of triangulation may be used to infer the position of a tagged animal without direct observation. Triangulation requires finding two or more bearings from different locations and plotting the

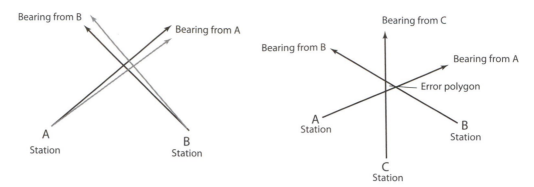

Figure 13.4 Receiver stations are used to establish bearings to the tagged animals. Using two stations (*left*) yields an estimated position for the animal where the two bearings intersect. If each bearing is off by a degree or two, then the estimated position changes dramatically. Using three stations (*right*) increases accuracy and reduces the size of the error polygon.

intersection of those bearings to determine one location point. This process is repeated to collect additional points over time. Each "location" point has an error polygon associated with it (Figure 13.4). The size and shape of this error polygon is determined by a number of factors, including

- the distance and angle between the bearings when they were taken,
- the distance from the receiver to the tagged animal,
- the beam width of the antenna, and
- the accuracy of the compass bearings.

Ideally, the two bearings should be taken close to the animal and at 90 degrees to each other (Figure 13.4), and a third bearing is usually taken from a third location. The polygon formed by the intersection of the three bearings should be relatively small; the center of the error polygon is considered the animal's position at that time. Obviously, if the researcher is alone and must move from one bearing point to another to take additional readings, then the readings will not be recorded at exactly the same time. If the animal is moving during the time the bearings are taken, then the error polygon will be larger. Consequently, triangulation studies typically rely on stationary base points that are permanently marked and numbered. The personnel at each station should take their bearings at exactly the same time and record them on standard data forms.

The accuracy of a radio location varies with habitat type and weather conditions. In mountainous areas, a common source of error is signal bounce, where a radio signal bounces off a ridge or mountain, resulting in huge errors. To reduce sources of error, researchers should ensure that the handheld antenna is matched to the frequency of the transmitters and held as high above the ground as possible (permanent stations may use poles to elevate the antennae). Alternatively, taking bearings from hilltops or ridgelines may improve the accuracy of the bearings. In addition, it is good practice to take as many bearings as possible, repeat bearings over short time intervals, and get as close as you can to the tagged animal without disturbing it.

Exercise 1: Locating Animals by Homing

After you arrive at your designated field site, your instructor will provide each team of students with a radio transmitter, a receiver unit, and antenna. Your instructor will familiarize you with the operation of the receiver unit before conducting the exercise. Make sure that the receiver is working properly and is receiving the signal from your transmitter. This exercise simulates how radio telemetry equipment is used to locate a tagged animal. One student is chosen to be the "animal" in the study, and the remaining members of the group are the "trackers."

The "animal" is given a working transmitter and asked to move 100 meters from the "trackers." Each member of the "trackers" should take turns listening for the signal while they scan a full 360-degree circle. In theory, the signal should be strongest when the antenna is pointing directly at the "animal." Each "tracker" should practice taking a compass bearing along the line of sight of the antenna's strongest signal. Bearings should be checked with the instructor for accuracy.

Once all "trackers" are confident that they can locate the strongest signal and take accurate compass bearings, the "animal" is asked to move out of sight and walk slowly to a specific location in the habitat (provided by the instructor). At the designated site, the "animal" places the transmitter on the ground or on a branch of a tree and returns to the "tracker" group. "Trackers" take turns locating the signal, taking a compass bearing, and moving a given number of paces (distance is set by the instructor). The exercise ends when the "trackers" have located the transmitter.

It is essential to set your compass declination before recording any bearings. Magnetic declination is the difference between true north and magnetic north. True north is fixed, but magnetic north (the direction the needle of a compass will point) changes over time. Magnetic north is currently some 450 miles from the true North Pole. Maps are drawn using true north because it does not change. Compasses use magnetic north. In North America, the line of true north and magnetic north are the same at the line of zero declination, which runs through western Lake Superior and across the western panhandle of Florida. West of the line of zero declination, an uncorrected compass will give a reading that is east of true north. The opposite is true if you are working east of the line of zero declination. To correct for magnetic declination, maps typically print the magnetic declination (deviation from true north) to the left of the scale bar (e.g., on a USGS 7.5' quadrangle map). Every time you go into the field, you should set your compass declination for the region. If you do not, you will not be traveling in the same direction as on the map. Each compass will differ slightly in how it is corrected for declination, and you should follow the manufacturer's instructions. For example, if the declination at your location is 16° E, you set the declination by rotating the graduated circle on the outside of the compass until it is 16° E from the indicator marker at the top of the compass. To make sure you have set your declination properly, orient your compass so that the north end of the needle is lined up with the 0° mark on the graduated circle. If you are located west of the line of zero declination, then the index pin or marker on your compass should be west of the 0° marker on the graduated circle (and vice versa if you are east of the line of zero declination).

Taking a compass bearing is relatively straightforward. A bearing is a measurement of direction and is given in one of two formats, an azimuth bearing or a quadrant bearing. Azimuth bearings use a 360° scale to indicate direction. An azimuth scale compass is numbered clockwise with north as 0°, east 90°, south 180°, and west 270° (e.g., a bearing of 32° would be northeast). Quadrant scale compasses are divided into four 90° quadrants. Azimuth scale compasses are preferred for telemetry studies.

If you are using a sighting compass with a diopter sight, you simply hold the compass up to your eye, point the marker on the compass in the direction you want, look through the sighting hole, and read off the bearing in degrees. These compasses are generally accurate to roughly ±0.5°. If you are using a regular baseplate compass, or one with a sighting mirror, then open the cover or mirror, position it at an angle of about 45° above the base, and align a landmark in the direction of your bearing (e.g., a tree) with the mirror line. The mirror's vertical line can be seen projected over the compass dial and the bearing read off the compass bezel.

Exercise 2: Locating Animals via Triangulation

In this exercise, you will set up two or more permanent tracking stations and use triangulation to locate the tagged "animal." Your instructor will position the student groups at the tracking stations. Each station is given a letter code. Station locations are determined using a handheld GPS unit and plotted on a topo map of the area. At each station, students mark out the four main compass directions on the ground where the antenna will be positioned.

If possible, it is helpful to fix the antenna to the top of a tall wooden pole with a 12-inch dowel set perpendicular to the pole (and parallel with the ground). This dowel is set so that it is exactly in line with the antenna set on the pole, and it acts as a general pointing device to indicate the general line of the signal.

With the stations set, students at each station use two-way radios to let the other stations know that they are ready to begin taking readings. The instructor will signal when to begin and the interval between readings (generally about 3-5 minutes apart). To take a bearing, slowly turn the antenna (and pole) in a 360° circle until the strongest signal is located. Take a compass bearing along the line of that signal and record it on data sheets (or in a field journal) along with the time and station code. In this case the tagged "animal" is allowed to move, so the bearings will change over time.

If the instructor does not have animals already tagged at the study site, one or more students can play the role of "animals" by carrying one of the radio collars as they slowly walk around the study area. Students acting as the "animals" should carry a handheld GPS unit and records their path as a set of waypoints at the same time intervals as the students taking bearings are recording the "trackers." In this way, the triangulation points can be compared with the actual GPS position of the "animal." At the end of the tracking session, students from each station share data to create a complete set of bearings for all stations at each time point.

Triangulating the Animal's Position

The basic principles of triangulation involve locating the receiver stations on a base map and drawing temporary bearing lines from each station. The place where the lines cross is the position (or error polygon) of the animal at the time the bearings were recorded (Figure 13.4). The steps are as follows:

1. Position a sheet of tracing paper over the topo map of the study site, and secure it to the map so that it doesn't move.

2. Carefully locate the receiver stations on the base topo map of the study area. Place a dot and the station's code letter at each location in pencil.

3. Place a transparent 360° compass circle (Figure 13.5) over the station so that the center of the compass circle is at the station's position.

4. Orient the compass circles so that N on the compass circle aligns with N on the map. Repeat for the other stations.

5. Beginning with the first set of bearings (e.g., those taken at the same time), use a straightedge to draw a faint pencil line that extends from the station along the bearing (e.g., for a bearing of 43° E, the pencil line would extend through the tick mark on the compass circle that corresponds to 43° east (Figure 13.5).

6. Repeat this process for the other stations using the bearings taken at the same time. Extend each line far enough from the station so that the lines all intersect.

7. At the intersection of all the lines, create a small solid circle at the center of the polygon and write the time next to that point (Figure 13.6).

8. Erase the temporary bearing lines used to create the point (do not erase the point or its corresponding time).

9. Repeat this process until all of the locations have been entered onto the tracing paper.

10. Follow the steps in Exercise 3 to estimate the area covered by the animal(s).

Figure 13.5 An example of a compass circle positioned over a station at + and a bearing of 43° NE.

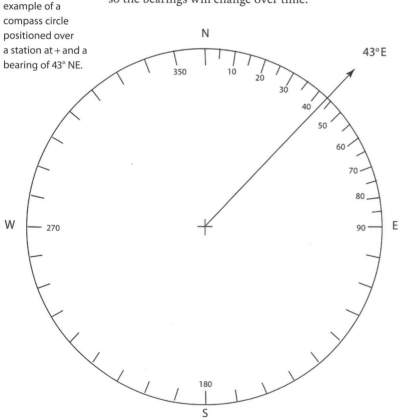

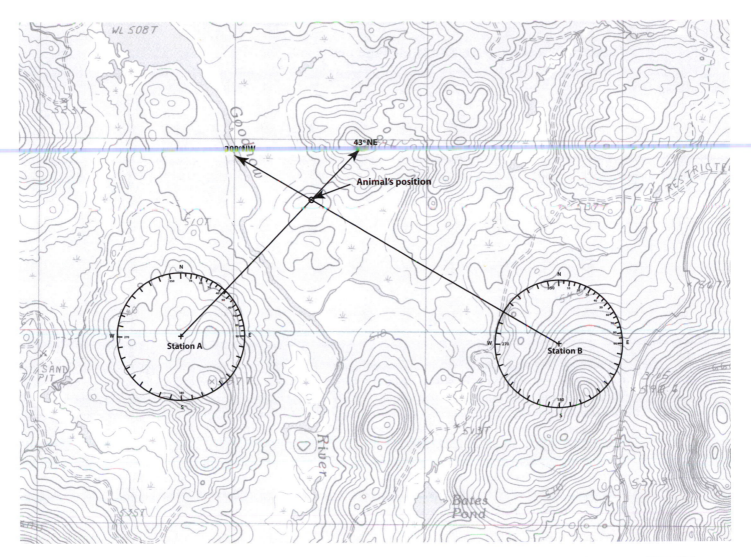

Animal's position

43° NE

Station A

Station B

Figure 13.6
Example of two compass bearings taken from two tracking stations at the same time. The intersection of the two lines represents the animal's estimated position at that time.

Exercise 3: Data Analysis: The Minimum Convex Polygon

Figure 13.7 Examples of a set of locations (circles) bounded by (**A**) a concave polygon with several indented regions (arrows) and (**B**) a minimum convex polygon.

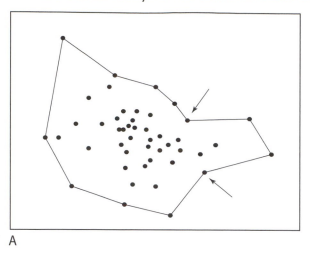

A

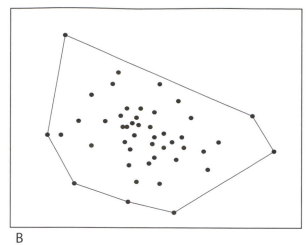

B

In this exercise, we will estimate the home range of the "animals" you tracked using minimum convex polygons (MCPs). A minimum convex polygon is one that contains within its perimeter all the line segments connecting any pair of its points (Figure 13.7). The polygon cannot have any indented regions on the perimeter; such a polygon would be said to be a concave polygon. To construct a minimum convex polygon, draw a pencil line around all the perimeter points to create a perimeter with no concave regions. This represents the potential home range of the animals tracked. However, it may not be an accurate representation of the actual home range used by the animal. To see why this is so, look at the location points in Figure 13.7. Notice that many of the points are clustered in a particular region of the home range. These points may represent areas of cover, places rich in food resources, or den sites. We will discuss how to deal with this in Exercise 4. For now, we will convert the MCP for our data into an area.

The formula for area of an MCP is

$$A = \frac{x_i(y_n - y_2) + \sum x_i(y_{i-1} - y_{i+1}) + x_n(y_{n-1} - y_1)}{2}$$

where x_i and y_i ($i = 1, 2, \ldots n$) are the coordinates for the locations.

This can be calculated by hand, but it is a difficult task with many location points. There are computer programs that will do the calculations for you, and we will discuss an example in Exercise 4. There are other ways to get a reasonable estimate of MCP area by hand. One approach involves plotting the MCP on poster board, cutting it out carefully, weighing the cutout MCP, and

comparing that weight to another cutout of a known area (e.g., 100 square meters or 1 square kilometer). Try this approach using a photocopy of your MCP and a square of known size (using the same scale as for the MCP).

A second approach uses a grid whose squares are of known size (Figure 13.8). The MCP is traced onto transparent paper, which is then laid on top of and secured to the grid. The area of one small square on the grid is determined using the scale on the map, and this area is multiplied by the total number of small squares that fall within the perimeter of the MCP. For example, if one small grid square is 1 square kilometer and there are 340 small squares that fall within the perimeter of the MCP, then the area of the MCP is 340 square kilometers. Try this approach using the grid paper provided by your instructor.

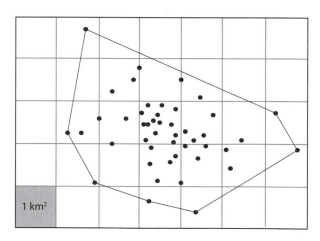

Figure 13.8 A minimum convex polygon of location points placed on top of a grid where one grid square is 1 km².

Exercise 4: Data Analysis Using sigloc in RStudio

In this exercise, we will use an R package called sigloc, which is short for "signal location" (Berg, 2015). Begin by launching RStudio and creating a New Project. In the Console window, enter the command for installing the sigloc package:

```
install.packages("sigloc")
```

Enter the Library command to create a library with dependent packages. There are two dependent packages (nleqslv and ellipse). Enter the following commands:

```
library(sigloc)
require(sigloc)
```

Load the data set called "bear," and take a look at what it contains:

```
data(bear)
View(bear)
```

The data set is a small subset of data collected by radio tracking a female back bear (*Ursus americanus*) in northern Minnesota. It includes columns for date, observer name, a grouping number (GID) for observations, time, UTM coordinates given as Easting and Northing, and an azimuth bearing. Now we want to convert the data into the desired format for plotting. Enter the following command:

```
bear <- as.receiver(bear)
```

Calculate the location of the places where the bearings intersect using

```
(cross <- findintersects(bear))
```

The output is a list of GIDs with *X* and *Y* values. Next we want to plot those intersecting bearings by using the plot command, as follows (this command prints the word "Easting" as the *x*-axis label and the word "Northing" as the *y*-axis label):

```
plot(cross, xlab="Easting",
ylab="Northing")
```

Make sure you have selected the tab Plots in the lower-right window. You should see the plot shown in Figure 13.9.

Enter the locate command below to estimate the transmitter locations using maximum likelihood estimation (MLE). It will print a table of values and show that one point in group 4 is "bad."

```
(loc <- locate(bear))
```

Plot the positions, including the bad point, on the existing graph by using the plot command with badcolor=TRUE and add=TRUE as arguments:

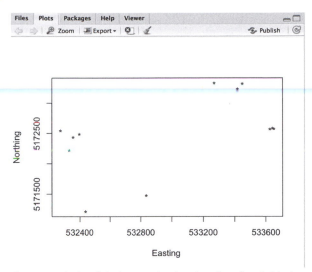

Figure 13.9 A plot of the intersection bearings for a female black bear data set in "sigloc" for R.

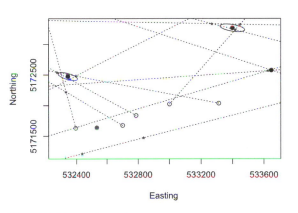

Figure 13.10 A plot of locations, error ellipses, and azimuth bearings for a female black bear using "sigloc" in R.

```
plot(loc, badcolor=TRUE, add=TRUE)
```

Finally, add the azimuth bearings to the existing plot:

```
plot(bear, bearings=TRUE, add=TRUE)
```

Your final graph should look like the one in Figure 13.10.

Exercise 5: Using adehabitatHR in RStudio

In this exercise, you will import a data set of latitude and longitude coordinates and use the R package adehabitatHR to plot minimum convex polygons and estimate the home range size (Calenge, 2006). Begin by opening RStudio (clear the Console area if needed). At the > prompt enter the following commands:

```
install.packages("adehabitatHR")
library(adehabitatHR)
library(sp)
require(adehabitatHR)
```

These two commands install the adehabitatHR package and its dependencies. Next we want to add some data. Download the file called "animal.csv" from the Johns Hopkins University Press website that accompanies this manual (jhupbooks.press.jhu.edu/content/mammalogy-techniques-lab-manual). To read a .csv file from Excel file into R, use the command read.csv(file.choose(). This allows you to choose a specific file from a location on your computer and call it "animal." Run the command below, and you will see a popup window; browse to locate the correct file and upload it.

```
animal = read.csv(file.choose())
```

Now we need to create x and y variables. We can extract the first column of data as x coordinates and the second column of data as the y coordinates.

```
x <- animal[,'x']
y <- animal[, 'y']
```

Check it by plotting it with:

```
plot(x,y)
```

You should see the locations plotted as a scatted plot with *x* and *y* for axes (Figure 13.11).

Next we need to convert the *x* and *y* values to a data. frame, which we will call "xy." Run the following command:

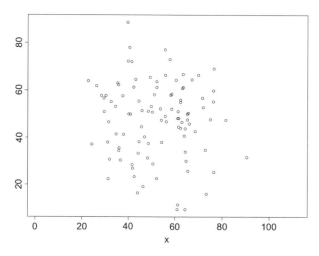

Figure 13.11 A scatter plot for the "animal.csv" data showing the positions of radio tracked animals using adehabitatHR.

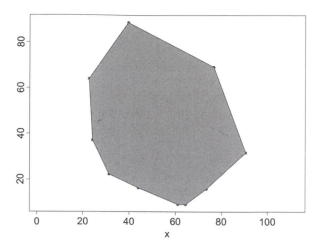

Figure 13.12 A 100% minimum convex polygon of the "animal.csv" data in adehabitatHR.

```
xy <- data.frame(x, y, 'id' = rep('a', 100))
```

Next we want to generate a set of minimum convex polygons around these animal locations. To do this we use the "coordinates" command. Run the three commands listed below:

```
coordinates(xy) <- xy[c('x', 'y')]
mcp_est100 <- mcp(xy[, 3], percent=100)
plot(mcp_est100, add=TRUE, col='orange')
```

These three commands generate a minimum convex polygon surrounding 100% of the locations and color it orange (Figure 13.12).

Suppose you want to also view the minimum convex polygons for 90% and 70% of the points. This can be useful to get a sense of where the animal spends most of its time. To do this, simply run the following commands:

```
mcp_est90 <- mcp(xy[, 3], percent=90)
mcp_est70 <- mcp(xy[, 3], percent=70)
plot(mcp_est90, add=TRUE, col='green')
plot(mcp_est70, add=TRUE, col='yellow')
```

Now the 90% MCP is colored in green, and the 70% MCP is colored yellow. Unfortunately, the colors are obscuring the data points. To change that, enter the command below, which adds red dots for each animal location on top of the colored MCPs:

```
plot(xy, cex=.75, pch=19, col= 'red',
add=TRUE)
```

If you do not like the color or shape of the points, you can change them by using some of the many features in the plot. For example, change cex=.9 and pch=25 using the command below:

```
plot(xy, cex=.9, pch=25, col= 'red',
add=TRUE)
```

You graph should now look like Figure 13.13. We still do not know the area of the 100% MCP. To find the area, run

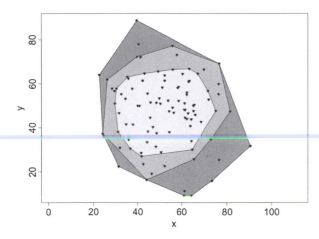

Figure 13.13 Minimum convex polygons for 100%, 90%, and 70% of the locations in "animals.csv."

the mcp command and ask for the units to be in kilometers and square kilometers:

```
mcp(xy, percent=100, unin=c("km"),
unout=c("km2"))
```

The resulting output data in the Console window should look like these lines:

```
Object of class "SpatialPolygonsDataFrame"
(package sp):
Number of SpatialPolygons: 1
Variables measured:
 id area
a a 3601.057
```

This says that the 100% MCP has an area of 3,601 square kilometers. Repeat this for the 90% and 70% MCPs.

14 GPS Tracking Using GPSVisualizer and MoveBank

Time Required
- One typical 3- to 4-hour lab period.

Learning Objectives
- Understand how the Global Positioning System (GPS) is used to track animals.
- Learn the format for GPS coordinates.
- Use Google Earth to analyze GPS data.
- Use GPSVisualizer online software to convert file types and plot GPS data.
- Access and analyze telemetry data sets from MoveBank.

Equipment Required
- Computers with access to the internet
- Data sets of GPS coordinates

Background

In radio telemetry, a tagged animal wears a radio transmitter that sends out a radio frequency signal that can be picked up by a researcher carrying a radio receiver tuned to that frequency (see Chapter 13). This gives the approximate direction to the animal, but only with triangulation from several locations can the animal's approximate position be determined. Radio telemetry is a relatively inexpensive way to track animals over relatively short distances. Researchers wishing to track large mammals over long distances (e.g., seasonal migrations) or from more remote sites now use GPS and/or satellite telemetry systems. GPS uses a series of 24 to 32 geostationary satellites in medium-Earth orbit that send signals to GPS devices on the ground. Using information from a minimum of four satellites, the GPS unit calculates the position on the ground (latitude, longitude, and altitude) plus the current time. GPS units can be attached to animals, and the location data can be stored on the GPS collar or transmitted to the researcher via satellite or cellular phone signals. This means that a researcher half a world away can track the movements of a tagged animal in near real-time (Handcock et al., 2009; Hebblewhite and Haydon, 2010; Tomkiewicz et al., 2010).

A GPS-enabled tag carried by an animal can record and store location data at a predetermined interval (e.g., every hour or twice a day). These data are stored in an onboard microprocessor, and the data are recovered when the animal is recaptured and the collar removed. Alternatively, GPS collars can be retrieved via remote-release systems that detach the collar automatically. Remote-release collars send out a radio signal upon release and are located in the field using radio telemetry.

More expensive GPS tags are coupled with satellite transmitters that relay the animal's position information to a satellite and from there to a central data center on the ground (Figure 14.1). For example, an animal with such a collar would receive a signal from GPS satellites, calculate its three-dimensional position on the Earth, and transmit the location to an Advanced Research and Global Observation Satellite (Argos), which would then relay that information to a ground-based data-processing center for transmission to the researcher. The researcher plots the animal's location on a map in near real-time using geographic information system (GIS) software.

In the nearly 50 years since the Craigheads (Craighead, 1982; Craighead et al., 1995) first radio tracked grizzly bears in Yellowstone National Park, there has been a radical change in how ecologists study the distribution and movement patterns of mammals (both terrestrial and marine). As Hebblewhite and Haydon (2010) point out, "Today, ecologists sitting at their desk can check the movements of even the most difficult to study species such as GPS radio-collared wildebeest (*Connochaetes taurinus*) or Argos-tagged bluefin tuna (*Thunnus thynnus*) on Google Earth, as they check their morning email with minute-by-minute data streams."

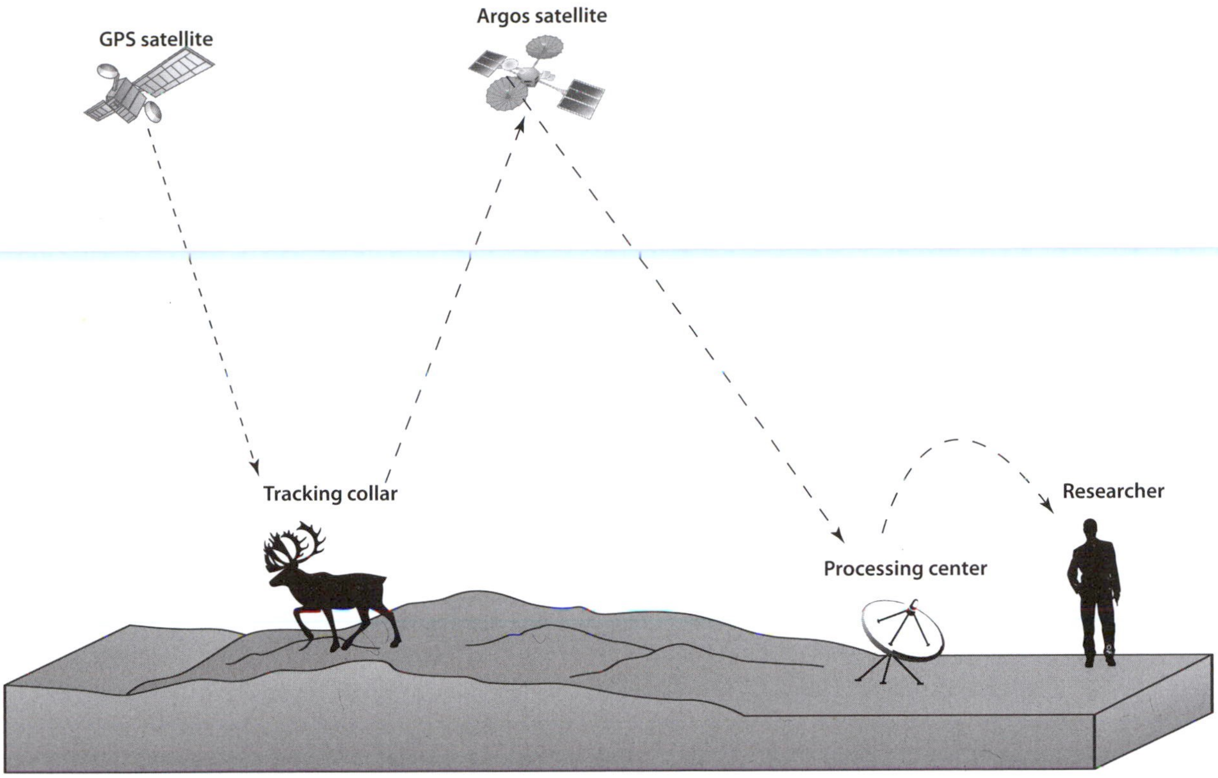

Figure 14.1
A diagram of the GPS-Argos system for remotely tracking large animals.

Exercise 1: Tracking Grizzly Bears with GPSVisualizer

This exercise uses GPS data for grizzly bears from North America. The data are in spreadsheet format. The data file—generously provided by Dr. Rick Mace at Montana Fish, Wildlife, and Parks—can be downloaded at the web page that accompanies this text. To begin, open the file "Griz-June/Oct.xls" in Excel. Look at how the data are formatted (Figure 14.2). The data are separated out by month, with the June data in one worksheet and the October data in another (see the tabs at the bottom of the spreadsheet). The main data we are interested in consists of the latitude and longitude, which will allow us to fix the animal's position in space and time.

In order to display these coordinates visually, you will need to map it. To do this, you will take advantage of two web-based software programs: Google Earth and GPSVisualizer. Google Earth is a powerful tool for displaying data on satellite images of the entire world. However, Google Earth cannot read raw spreadsheet data (such as .xls files from Excel). The first step is to convert the raw spreadsheet data into a format that Google Earth can display.

1. Open your web browser, and navigate to the GPSVisualizer website at http://www.gpsvisualizer.com/.

2. In the Get Started Now box, click the Choose File button and locate the file "Griz-June/Oct.xls" on your computer. Select this data file, and choose Google Maps

	A	B	C	D	E	F	G	H
1	MONTH	DAY	YEAR	gmt	GMT_HR	DEAD_ALIV	lat	long
2	6	13	2007			alive	48.87068	-113.90928
3	6	14	2007	0.00154	0	alive	48.86902	-113.90689
4	6	15	2007	0.00076	0	alive	48.86997	-113.9002
5	6	15	2007	0.50076	0.5	alive	48.87483	-113.89709
6	6	15	2007	0.75044	0.7	alive	48.86923	-113.90634
7	6	16	2007	0.00096	0	alive	48.87525	-113.90548
8	6	16	2007	0.50076	0.5	alive	48.86898	-113.89576
9	6	16	2007	0.75043	0.7	alive	48.87457	-113.89742
10	6	17	2007	0.0011	0	alive	48.87528	-113.90377
11	6	17	2007	0.75064	0.7	alive	48.87909	-113.90042
12	6	18	2007	0.00075	0	alive	48.87535	-113.90215

Figure 14.2 Format of the data in the grizzly data set in Excel.

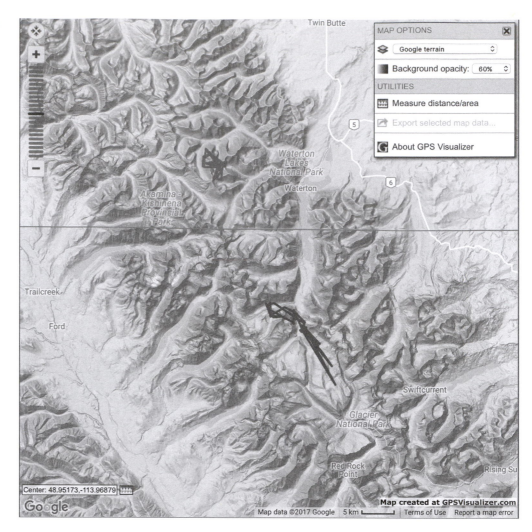

Figure 14.3 Google Maps display of the tracking data for June (*bottom*) and October (*top*) overlaid on a terrain map of the region. The horizontal line in the middle is the US-Canada border. *GPS Visualizer (www.gpsvisualizer.com).*

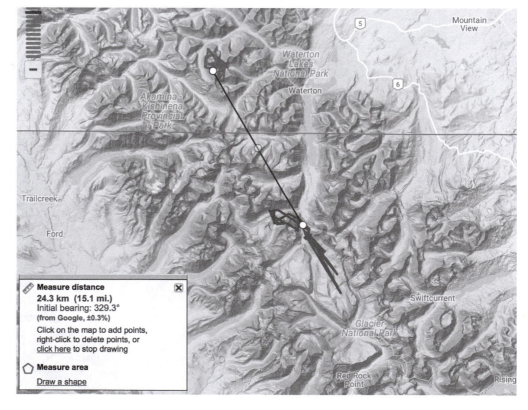

Figure 14.4 Google Map display with the distance between the June data and the October data shown in the box as 24.3 km. *GPS Visualizer (www.gpsvisualizer.com).*

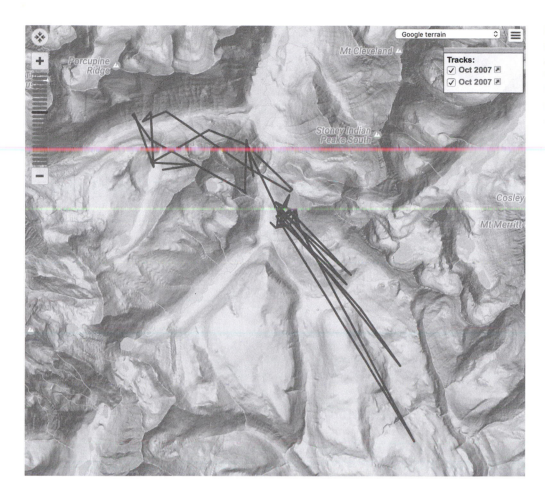

Figure 14.5
A close-up of the June data set in Glacier National Park. *GPS Visualizer (www.gpsvisualizer .com).*

as the output format. Click the Map It button to display the data on a Google Earth map.

3. In the upper-right corner of the map, you will see (1) a pull-down menu for the type of map displayed, (2) a small box with green lines that represents the map utilities, and (3) the Tracks box. Explore these options by clicking on the Map Utilities box and selecting Google terrain as the map type and 60% as the opacity. You will see a page like the one in Figure 14.3, except your map will be in color.

4. In the Map Utilities, click on the Measure distance/ area tab. You can now click on the center of the blue data set in Waterton Lakes National Park, Canada, and again on the center of the red data set in Glacier National Park, USA, to get the distance between the two data sets (Figure 14.4).

5. In the Tracks box, you will see the two data sets listed (note that although the Tracks box indicates that both data sets have the same name, the blue set is from October 2007 and the red set is from June 2007). Click on the small boxed arrow to the right of the red Oct 2007 data to zoom to that location (Figure 14.5).

6. You can download a copy of your map by clicking on the word Download in the sentence above the map (under the heading Google Maps output).

7. Click on the link to the right of your map called Return to the GPSV Map Form. This takes you to a de-

tailed form where you can change how the data looks and extract additional information, such as the location of each track point. Fill out this form as shown in Figure 14.6., On the upper right side, click Choose File and browse to the Excel file Griz JuneOct.xls again to upload it into this form. Click the Draw the Map button. You can now see each waypoint as a dot. Zoom in to the blue data set (October), and click on one of the waypoint dots. A pop-up box is displayed with the latitude and longitude of that point (Figure 14.7).

8. You can also convert your map to a format Google Earth can open by reloading GPSVisualizer, uploading the Excel file again, and clicking on the Google Earth KML option from the options list (Figure 14.8).

9. Fill in the form for your Google Earth KML map exactly as shown in Figure 14.9. Remember to upload the Excel file again in the Choose File area and click the Create KML file button. Click on the file that ends in map. kmz to download your map. Now open this file in Google Earth (File > Open). The data are now available for viewing and manipulating.

10. Zoom in on the blue data set to see how the grizzly bear used the terrain in October. Adjust your observation point using the navigation tools in the upper-right corner of the Google Earth display. Notice that this female spent June in Glacier National Park in the US and October in Waterton Lakes National Park, Canada.

Figure 14.6 The Google Earth KLM window in GPS-Visualizer. Set the parameters to match those shown here. *GPS Visualizer (www .gpsvisualizer.com).*

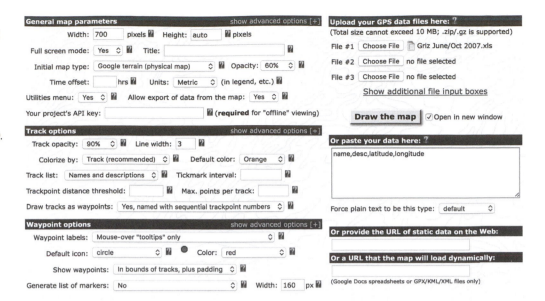

Figure 14.7 Close-up of the Google Maps showing the data for trackpoint 1. The data now include latitude and longitude for each trackpoint. *GPS Visualizer (www .gpsvisualizer.com).*

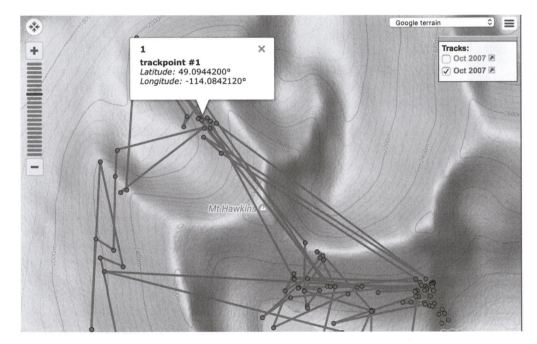

Figure 14.8 The options window in GPSVisualizer. *GPS Visualizer (www .gpsvisualizer.com).*

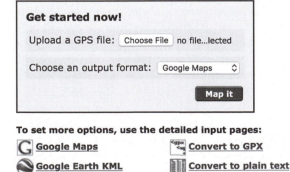

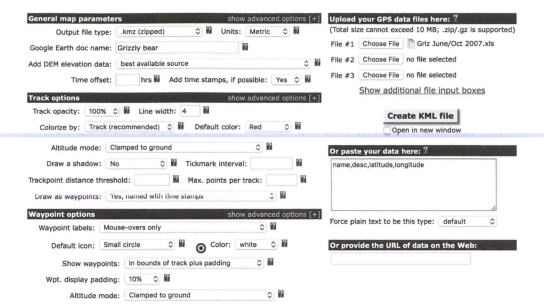

General map parameters   show advanced options [+]

Output file type: .kmz (zipped)   Units: Metric

Google Earth doc name: Grizzly bear

Add DEM elevation data: best available source

Time offset: ___ hrs   Add time stamps, if possible: Yes

Track options   show advanced options [+]

Track opacity: 100%   Line width: 4

Colorize by: Track (recommended)   Default color: Red

Altitude mode: Clamped to ground

Draw a shadow: No   Tickmark interval: ___

Trackpoint distance threshold: ___   Max. points per track: ___

Draw as waypoints: Yes, named with time stamps

Waypoint options   show advanced options [+]

Waypoint labels: Mouse-overs only

Default icon: Small circle   Color: white

Show waypoints: In bounds of track plus padding

Wpt. display padding: 10%

Altitude mode: Clamped to ground

Upload your GPS data files here:
(Total size cannot exceed 10 MB; .zip/.gz is supported)

File #1 Choose File   Griz June/Oct 2007.xls

File #2 Choose File   no file selected

File #3 Choose File   no file selected

Show additional file input boxes

Create KML file
☐ Open in new window

Or paste your data here:
name,desc,latitude,longitude

Force plain text to be this type: default

Or provide the URL of data on the Web:

Figure 14.9 The Google Earth KLM window in GPS-Visualizer. Set the parameters to match those shown here. *GPS Visualizer (www .gpsvisualizer.com).*

General parameters   show advanced options [+]

OUTPUT FORMAT: PNG -- better than JPEG for profiles

Profile width: 590 pixels   Profile height: 340 pixels

Profile margin: 80 pixels (added to width & height on all sides)

Units: Metric   Title: Grizzly 2007

X-Axis: Distance   X divisions: auto

X min.: ___   X max.: ___

Y-Axis: Elevation   Y divisions: auto

Y min.: ___   Y max.: ___

Add DEM elevation data: best available source

Track/line options   show advanced options [+]

Drawing mode: Lines   Line width: 2

Colorize by: Elevation   Min: ___ Max: ___

Cumulative distance: Yes   Track dividers: Yes

Include waypoints, if possible: No

Upload your GPS data files here:
(Total size of all files cannot exceed 10 MB)

File #1 Choose File   Griz June/Oct 2007.xls

File #2 Choose File   no file selected

File #3 Choose File   no file selected

Show additional file input boxes

Draw the profile
☑ Open in new window

Or type/paste your data here:
latitude,longitude,altitude

Force plain text to be this type: default

Or provide the URL of data on the Web:

Or draw a cross-section between two points:

Point 1: ___   Point 2: ___

Number of trackpoints to interpolate: ___

Figure 14.10 The draw profile entry window in GPS-Visualizer. Fill it out exactly as shown here. *GPS Visualizer (www .gpsvisualizer.com).*

11. One of the things we can see from this map is that the area is mountainous. You might want to know at what elevations the female grizzly spent most of her time and whether the elevation changed with the seasons. You can add elevation information to the GPS points even though they were not part of the original spreadsheet data. To do this, we again use GPSVisualizer.

12. Return to GPSVisualizer and click on Google Earth KLM again. Fill in the fields exactly as shown in Figure 14.9 (including uploading the Excel file again using Choose File button in the upper right). In the general map parameters section, set the Add DEM elevation data to best available source. You may change a few settings such as "default color" and the icon type for your waypoints if you like, but leave the rest of the settings as shown in Figure 14.9. Click the Create KLM file button, download the resulting file, and open it in Google Earth.

13. After Google Earth reloads the data, the elevation data are attached to each waypoint. Drag your mouse over any waypoint on the map, and/or click on one waypoint to call up a balloon with the elevation data for that point. Export the elevation data to your spreadsheet so you can calculate mean elevations and other parameters.

14. Return to the GPSVisualizer main page, and click on the button called Profiles (elevation, etc.). It will open another data entry screen. Fill in the data boxes as in Figure 14.10 (including uploading the Excel file again using Choose File button in the upper right).

15. Click the Draw the Profile button at the bottom right. You will see something like the plot in Figure 14.11.

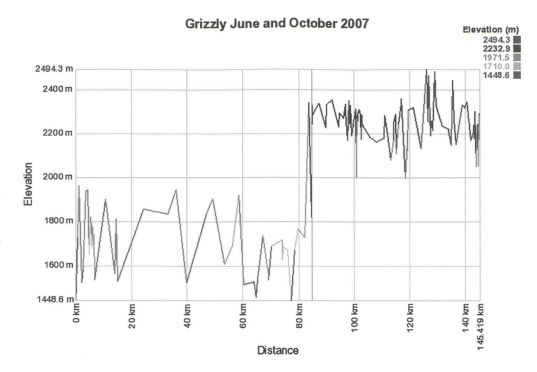

Figure 14.11 A plot of the elevation versus distance traveled by the female grizzly in June and in October 2007. Notice that there is a distinct break at about 80 km. This break represents the jump from June data to October data. *GPS Visualizer (www.gpsvisualizer.com).*

This plot shows the altitude of the grizzly for June and October. Notice that there is a sharp increase in altitude after the animal traveled approximately 80 kilometers. This is an artifact produced because the July to September data were deleted. Download your plot data by clicking the green download link in the text above the plot.

16. Suppose you want to store that elevation data with your original latitude and longitude data. To add elevation and distance data to your original spreadsheet file, open GPSVisualizer and click on Convert A File in the list of options along the top menu. You will get a gray data entry box. Fill it out by uploading the Excel file again with the Choose File button, checking the boxes for speed and distance, and setting the Add DEM elevation data drop-down list to best available source.

17. Now click the Convert button to get a .txt file. The display shows a box of what the data look like now. It may look a bit jumbled because the data are too wide for the display box. Click the green link called "Click to download . . . txt" to download and save your text file to your desktop. Import this .txt data file into Excel or another spreadsheet program: Open Excel, go to File>Import, select the check box next to Text file, and click the Import button. Locate the text file you just downloaded, and click the Get Data button. In the Text Wizard pop-up, click the Next button and then the Finish button. Click the check box for New Sheet to create a new spreadsheet. Your spreadsheet now has the altitude of each waypoint and the distance between consecutive waypoints.

One final note: you can create a tour of your waypoints (similar to a flyover). You can do this by selecting the appropriate data in "My Places" (on the left-hand side of the Google Earth interface). To do this, you need to find the appropriate line (path) in "My Places," select the appropriate line in the "Places panel," and click the Play Tour button (located at the bottom of the "Places panel"; it is a folder icon).

The tour begins playing in the 3D viewer, and the tour controls appear in the bottom-left corner of the 3D viewer. To pause or resume the tour, click the Pause/Play button. To fast-forward or rewind on the tour, click the arrow buttons (press these repeatedly to accelerate back or forward). To replay the tour, click the Repeat button. Use the tour slider to move to any part of the tour. These controls disappear if the tour is inactive for a period of time, but you can make them reappear by moving the cursor over the bottom-left corner of the 3D window.

Based on your work with the grizzly data, consider the following questions:

- What is the average rate of movement per day for each month for this female grizzly?
- What is the average elevation change per day in each month? What might explain the differences in elevation?
- What is the maximum elevation change and maximum distance moved in one day (for each month)?
- Do grizzly bears prefer dense forest or more open areas? To answer, you must zoom in on data points in Google Earth and inspect the habitat.
- Do grizzly bears prefer north- or south-facing slopes?
- What is the distance to the nearest road, town, water? You will have to click on the check boxes in the layers panel in Google Earth to see roads and other features added to your map. Compare distances in June (just after arousal from hibernation) with those in October (just before entering hibernation).

Exercise 2: Exploring MoveBank Data

In this exercise, you will use an online database of telemetry information to obtain and display GPS data for African buffalo in Kruger National Park, South Africa (Getz et al., 2007; Cross et al., 2016). The data set can be found online at MoveBank (http://www.movebank.org).

MoveBank is a community of researchers who have uploaded their personal data sets to a database to allow broader access to tracking data. Users can display or analyze these data using MoveBank's website. Individual researchers can choose to make their data available to others or keep it private.

Register now as a MoveBank user, and log in again using the username and password you have just created. You may use the data sets in MoveBank, but you may not publish the results without permission from the original owner (collector) of the data set. Read over the terms of use and copyright information.

1. Open the MoveBank page on your browser at http://www.movebank.org. Click on the Tracking Data map on the right to see a larger map with all the data sets displayed (Figure 14.12).

2. The main search page displays a data-searching panel on the left and a map on the right. The locations of the data sets in the database are displayed on the map with gray or green dots. The map can be set to display as a basic "map" or as a satellite image map by clicking on the buttons in the upper-left side of the map. Try switching between map settings, but return to "Map" before continuing with this exercise.

3. Using the data-searching panel on the left, enter the word `Mammal` in the top search panel to display only data sets where mammals were tracked. Scroll down the list of data sets until you locate the listing for "Kruger African Buffalo, GPS tracking, South Africa."

4. Click in the small check box immediately to the left of the name of the data set to select these data.

5. Click on the small icon on the right side of the data set's name; it looks like a small box with an *i* inside. This opens a pop-up screen with three choices.

6. Click on the choice Open in Studies Page. This will open a new page that provides a summary of the contents of the data set you will work with. The study details include the researcher, his/her address, the license terms, the number of tagged animals, the number of locations, and other relevant information. Notice that this is a very large data set.

Figure 14.12 MoveBank's main search page showing a map of all the data sets and a search panel on the left.

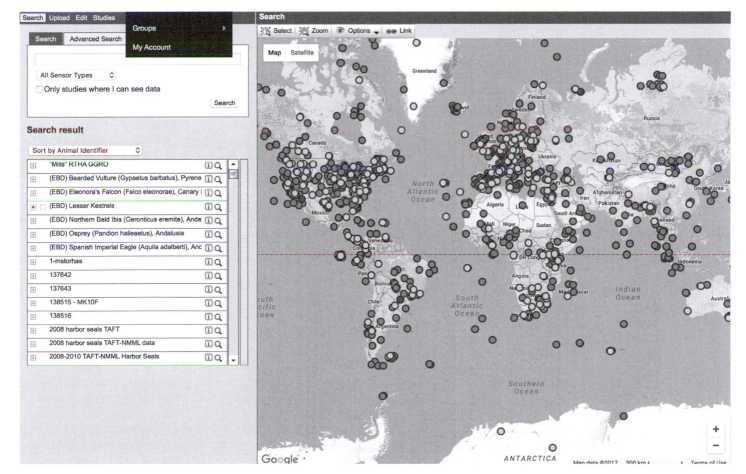

Figure 14.13 A terrain map of the African buffalo locations in Kruger National Park, South Africa.

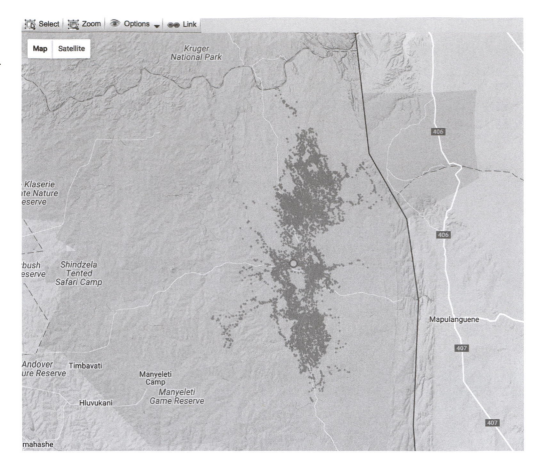

7. Use your browser's Back button to go back to the main data search page.

8. Click on the small magnifying glass icon next to the Kruger African Buffalo data listing in the search panel (Figure 14.12). This displays a terrain map zoomed in to the region in South Africa where the data can be seen as a series of pink patches (actually pink points; Figure 14.13).

9. You can use the + and − buttons on the lower right of the map to zoom in or out. This map shows two distinct herds of African buffalo, one to the north and one to the south. The boundary of Kruger National Park is shown in green in the terrain map, so you can see whether the herds are located within the park or sometimes cross park boundaries. Note also how the buffalo move relative to large rivers and roads in the area.

10. Click on the Options button at the top left of the map. You will see a drop-down menu. Click the box for Draw Lines for Selected Animals, and uncheck the box for Draw Lines for Highlighted Animals. Then click Close to close the drop-down menu.

11. Use the map zoom features to zoom in on the two buffalo herds. You can do this by clicking the + button several times (and re-centering your map), or you can drag the zoom slide bar up until it is in the fourth position down from the top +. You will now see many tiny pink dots representing the GPS locations for one buffalo at a specific time point. You should also see path lines be-

tween points with arrows pointing in the direction of movement.

12. Click on one path line (zoom in further if needed). This displays the points for one of the six African buffalo in blue. Now you can see how this individual moved relative to the other tagged buffalo in that herd.

13. Click back and forth between the Satellite and Terrain views to see how the herd moves relative to geographic features (e.g., steep ridges, open savanna, water sources).

14. Click on the small *i* icon on the right-hand side of the data set listing to display a pop-up of choices for what to do with the data. Click on Download Search Result. In the pop-up window that appears, click on Csv, Add UTM Coordinates, and Add Study Local Time, and then click Download. You can now explore this data set in Excel.

15. Click on the small *i* icon again to display a pop-up of choices for what to do with the data. Click on Download Search Result. In the pop-up window that appears, click on GoogleEarth (Tracks) and then on Download. Open the resulting .kmz file in Google Earth (as described in Exercise 1), or save the .kmz file to your computer for later use.

16. Google Earth displays the African buffalo data set.

17. Select a second data set to explore. Use the features described above to explore the animal's movements. For

example, you might use the Indiana bat data to explore how these bats leave their hibernacula, or explore how urban coyotes use suburban and urban environments at night.

18. Write a detailed description of how this animal uses its habitat. Some of the questions you should consider in your write-up include the following:

- What type of terrain is preferred?
- What type of cover is used?
- Do movements change with time of day or night?

- When is the animal most active?
- How far do the animals typically travel in a day?
- What altitude do they prefer?
- Do they avoid human-modified landscapes? If so, how?
- How do their movement patterns relate to water sources?

For some of these questions you may need to export the data to Google Earth or as a .csv file that you open in a spreadsheet.

15 Recording and Analyzing Mammal Sounds

Time Required
- Approximately one 3- to 4-hour lab period for each exercise.

Learning Objectives
- Understand the role vocalizations play in animal communication.
- Understand the importance of alarm calls in animal societies.
- Learn how to record animal vocalizations in the field.
- Understand how sounds are displayed visually in a spectrogram.
- Learn how to measure and manipulate audio information from a spectrogram.
- Apply knowledge of sound and communication to a field experiment.
- Conduct a playback experiment.

Equipment Required
- Binoculars
- Digital audio recorder/player (using SD memory cards)
- Bat detector (optional)
- Field microphone (shotgun or parabolic)
- Battery-powered field speakers
- Audio analysis software (Audacity)
- Data sheets and clipboards

Background

Mammals have an acute sense of hearing, and auditory communication is very important in most mammal species (Vaughan et al., 2015). Mammals use acoustic signals to maintain contact with other members of the herd or pack, to engage in courtship behaviors, to warn of danger, and to recognize individuals within a colony or social group (e.g., mother-infant communication). Primates and cetaceans in particular exhibit complex repertoires of acoustic signals (Goodall, 1986; Weilgart and Whitehead, 1997). With the advent of more sophisticated sound-recording equipment, researchers have demonstrated that mammalian auditory communication extends well above (ultrasonic) and well below (infrasonic) the human hearing range (see Vaughan et al., 2015).

Ultrasonic signals, or echolocation signals, have been known in bats and dolphins for decades. However, it is only in the past decade that researchers have discovered that some mammals communicate over long distances in the infrasonic range (e.g., below 20 hertz). African elephants produce intense, extremely low frequency vocalizations (14 to 24 hertz) that may carry over six miles (Payne et al., 1986). These varied infrasounds communicate locations of individuals or groups across vast distances; they also relay information on reproductive condition, social identity, and danger (Poole et al., 1988; McComb et al., 2003).

In another fascinating example, researchers discovered that some fossorial (living primarily underground) mammals communicate using a combination of low-frequency vocalizations and seismic signals (vibrations transmitted through the soil). For example, banner-tailed kangaroo rats (*Dipodomys spectabilis*) communicate by foot drumming (Randall, 1997; Randall and Matocq, 1997). Apparently, these kangaroo rats can recognize strangers and neighbors by their unique foot-drumming signatures (Randall, 1989).

Mammalogist know relatively little about acoustical communication in most wild mammals. Therefore, we can expect many exciting new discoveries in the future, and mammalian bioacoustics remains a promising area for research.

Equipment for Recording Sounds

Microphones

The microphone converts sound pressure into an electrical signal, which can be amplified, recorded, and analyzed. There are a number of designs on the market, and choosing the right microphone is critical. There are five main factors to consider:

- Type of transducer
- Efficiency (or sensitivity)
- Self-noise (its intrinsic noise)

- Frequency response (range of frequencies received)
- Directionality (polar pattern)

Microphones are typically classified by their transducer design (e.g., condenser or dynamic) and by their directional characteristics. Dynamic microphones are very reliable, more resistant to moisture, and do not require powering. However, they are not as sensitive as condenser microphones.

Condenser microphones have extended frequency response but require powering (usually via 48-volt phantom powering from the recording unit). An electret condenser microphone is a relatively new type of miniature microphone found in most consumer electronics today. Some of these microphones rival traditional condenser microphones in signal quality yet are small and inexpensive.

Another important consideration for field recording is directionality. In bioacoustic recordings, directional microphones (e.g., shotgun microphones) and parabolic microphones are most useful. Directional and parabolic microphones collect sound coming directly in front of the microphone and thereby reduce ambient noise (e.g., wind and human activities). Wind noise is reduced by adding special windshields to shotgun microphones. Parabolic microphones use a dish-shaped parabola to focus incoming sounds on a focal point where a microphone is positioned. Parabolic microphones are somewhat limited by the sound's wavelength. A low-frequency sound of approximately 100 Hz would require a parabolic dish three meters in diameter (not a practical field option).

Self-noise is also an important consideration for field recording. The best (quietest) have a self-noise of 5 to 10 decibels (dB), compared with the average microphone's 20 to 25 dB. Ideally, one would choose a directional microphone with the highest sensitivity, lowest self-noise, and widest frequency range.

Hydrophones are special transducers that acquire underwater sounds (pressure waves) and convert them into an electric signal. Hydrophones use piezoelectric elements that produce an electrical current when compressed by a sound wave. They are usually omnidirectional and have a frequency response from 2–10 Hz to more than 100 kHz.

Digital Audio Recorders

Audio recorders can be analog or digital. They allow recording an electrical signal generated by a microphone or hydrophone. Traditional cassette and open reel tape recorders (analog) degrade the signals by adding hiss or other types of distortion. Digital recorders eliminate these problems and record highly accurate, low-noise signals. Additionally, digital audio files can be easily stored, searched, and processed using computers. Thus, digital audio recorders are preferred by wildlife researchers (analog systems will not be described here).

Digital audiotape (DAT) recorders can still be found on the market but are being rapidly replaced by solid-state recording units. DAT has a frequency response of 10 Hz–22 kHz and cannot be used for ultrasonic signals. MiniDisc (MD) recorders are still available, but they use compression algorithms that degrade the sound (they often delete audio information outside the range of human hearing). The current state-of-the-art in digital recorders are units that record on internal hard disks or on removable memory cards (e.g., SD memory cards). These units can record in compressed or uncompressed formats depending on the duration or quality needs of the researcher. Another advantage is that some models can record up to 192 kHz, although 96 kHz is more common. Most are pocket-sized and have microphone preamplifiers. For a comparison of the recorders, visit the web pages of the Wildlife Sound Recording Society at https://www.wildlife-sound.org.

Infrasonic and ultrasonic recorders are specialized devices capable of recording signals whose frequencies are lower or higher than those audible to humans. Ultrasound (echolocation) recordings can be played back at slower speed to make them audible to humans. Devices capable of detecting and recording ultrasound exist to study echolocation in bats (e.g., bat detectors). Three main types of bat detectors are available:

- Heterodyne frequency-shifting detectors
- Frequency division detectors
- Time expansion detectors

Heterodyne detectors shift a small frequency range (several kilohertz) down to the human audible range. Unfortunately, anything outside that frequency range is lost. Frequency division (FD) bat detectors convert the bat's call into a square wave and then divide that by 10 to generate a new square wave. FD detectors cover a wide frequency range, but square waves sound harsh and contain unwanted harmonics. Heterodyne and FD systems allow recording of ultrasonic signals transformed to audible signals, not the original ultrasonic signal. The time expansion detector (TED) is the most accurate. All the information of the original signal is retained. The bat's ultrasonic signal is digitally sampled at high speed, stored to memory, and replayed at one-tenth the rate to make it audible (and recordable). For additional details, see Ahlén and Baagøe (1999) and the website http://www.bats.org.uk/pages/bat_detectors.html.

Overall, if you plan to record mammal sounds in the field, you will need a digital recorder with the following characteristics:

- Long battery life and ability to use normal or rechargeable AA or AAA batteries
- AC power supply capable
- Portable case size/button size but easy to use in the field

- 1 GB memory capacity (or removable memory cards)
- Manual control of record level
- Bright, easy-to-see record-level meters
- XLR microphone connectors for attaching external mics
- ¼-inch phono jack for attaching headphones
- +48V phantom power (for external mics)
- A sample rate of at least 48 kHz and a 24-bit depth
- An internal speaker or the capacity to connect an external speaker to the recorder for playbacks

Software for Analyzing Sounds

Sound analysis software allows users to

- graphically display acoustic signals
- understand and measure their structure, and
- correlate it to observed species, behaviors, and situations.

This exercise assumes that readers are familiar with basic terms such as frequency, amplitude, wavelength, hertz, and decibels. If not, refer to Vaughan et al. (2015) before proceeding.

Audacity, a free (open-source) software, can be used to record and analyze animal sounds (http://audacity .sourceforge.net/). Another free sound analysis program is Raven Lite (Raven Pro is a full feature commercial version) produced by the Cornell University Lab of Ornithology (http://www.birds.cornell.edu/brp/raven /RavenVersions.html#RavenLite). The discussion that follows will use Audacity.

Audacity produces three types of graphs commonly used to visualize sounds. First, oscillograms or waveforms display sound intensity fluctuations over time (seconds). Here, time is shown on the x-axis, and sound volume (amplitude) is reflected in the height of the spikes on the y-axis (Figure 15.1). This is the default graphical display in Audacity.

Second, sonograms give information on the frequency (pitch) of the sound. The x-axis again represents time in seconds, and the y-axis displays frequency, with low-frequency sound near the bottom (Figure 15.2). The frequency of the lion growl in Figure 15.3 ranges from 80 Hz to 5,500 Hz (or 5.5 kHz). In a sonogram, the amplitude (volume) is illustrated by the intensity of the colors. In black and white, the highest-intensity sounds are black and the lower-amplitude sounds are in lighter shades of gray. In a color sonogram, amplitude variations are displayed in different colors (set by the user).

Third, power spectrograms display frequency versus amplitude in decibels (dB) summed over the length of the sound selected (or over the entire song). Frequency is displayed on the x-axis, and amplitude (volume in dB) on the y-axis (Figure 15.3). These plots are generated using fast Fourier transform (FFT) algorithms. It shows the sound energy present at each frequency within the sound file. For this lion growl, the highest amplitude

Figure 15.1 A waveform (oscillogram) display of a lion roar and growl in Audacity.

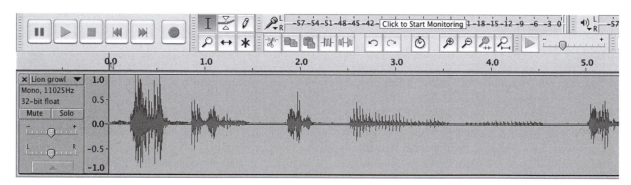

Figure 15.2 A sonogram of a lion growl in Audacity.

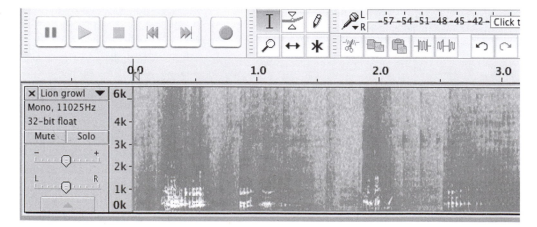

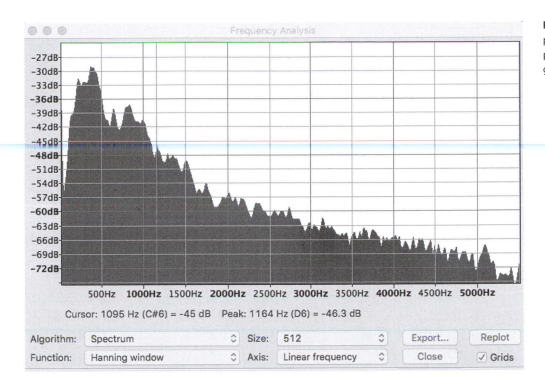

Figure 15.3 A power spectrum plot for a lion growl in Audacity.

Figure 15.4
A sonogram of a *Rhinolophus blasii* bat. Note the frequency range extends up to 125 kHz or well above the human hearing range.

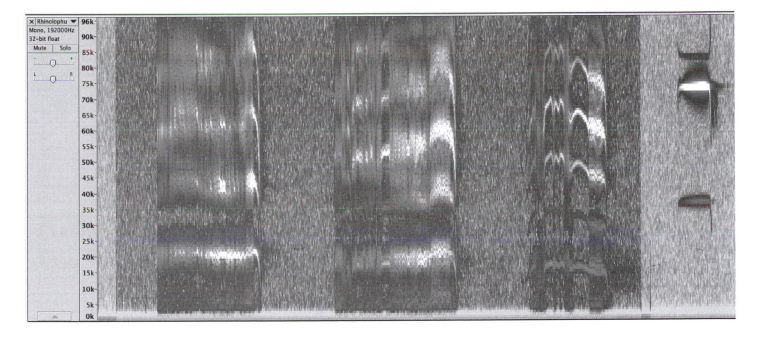

(loudest) sounds occur at frequencies around 390 Hz. This is not surprising given that lions have a very deep, throaty growl. For animal sounds, the most useful display is the frequency-time display, with intensity coded in greyscale or a color scale. This kind of analysis is usually called a sonogram or spectrogram (Figure 15.2). These graphic displays of audio information allow researchers to compare signals from different species, different individuals, or the same individual in different behavioral contexts. Sonograms (or spectrograms) may show infrasounds, like those emitted by elephants, as well as ultrasounds emitted by echolocating bats (Figure 15.4).

Spectrograms may reveal features, like fast frequency or amplitude modulations that we cannot hear even if they lie within our hearing frequency limits (30 Hz–16 kHz). Spectrograms are widely used to show the features of animal calls or vocalizations.

Interpreting a Sonogram

A tone with a constant frequency (such as a single note on a piano) will appear as a horizontal line with its vertical position depending on the frequency of the tone. A tone with increasing or decreasing frequency will appear

Figure 15.5 Two contrasting bat echolocation signals. (**A**) A constant frequency (CF) signal from a bat in the genus *Rhinolophus,* and (**B**) a frequency modulated (FM) signal from a bat in the genus *Nyctalus.*

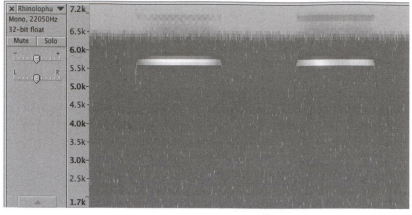

A

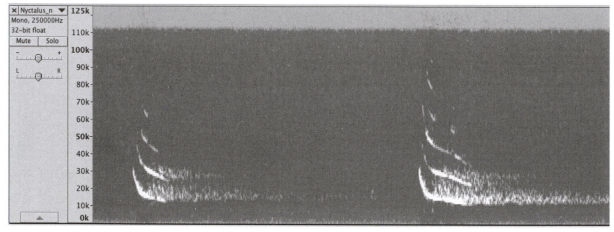

B

as an inclining or declining line (Figure 15.5). For example, in the *Rhinolophus* signal at the top in Figure 15.5, the frequency is held constant at about 95 kHz for 0.04 seconds (40 milliseconds); this is referred to as a constant frequency (CF) echolocation signal. In contrast, the signal from *Nyctalus,* at the bottom in Figure 15.5, shows echolocation pulses where the signal sweeps downward from a high of roughly 50 kHz to a low of 20 kHz over 0.02 seconds (20 milliseconds). *Nyctalus* also displays a signal that has multiple harmonics. The lowest tone is called the fundamental frequency, and tones that are multiples of this are called harmonics. For a detailed discussion of bat echolocation signals, readers are referred to Vaughan et al. (2015).

Exercise 1: Field Recording

In this exercise, you will record animal sounds in the field for later analysis. Your instructor will provide each group of students with a set of recording equipment and will review the operation of the specific equipment provided.

1. Assemble the entire recording system, and check the equipment in advance. Make a test recording, by speaking into the microphone, to make sure everything is connected properly.

2. Walk slowly down a trail near where, based on previous experience, you expect to encounter squirrels (or another mammal).

3. Try to avoid recording on windy days. If necessary, use your body to shield a shotgun mic from the wind. Wind speed is much less near the ground, so kneeling or holding the microphone as close to the ground as possible will help reduce wind noise. If you are using a parabolic mic, hold it so that the wind is coming from the back of the dish.

4. Get as close to the animal as you can without disturbing it. Halving the distance to the animal doubles the signal level reaching the microphone.

5. Point the microphone directly at the vocalizing animal (try to avoid obstructions between the mic and the animal).

6. Set the record level so that the loudest sounds register just into the red region (color may vary on different devices) of the VU meter (peak meter).

7. Record for at least one minute, or longer if the animal allows.

8. Minimize your body movements and other sounds.

9. Announce date, location, species, and other basic data at the end of each recording.

10. Listen through headphones to recordings you have just made to ensure they are recorded correctly.

11. Take turns making recordings so that every student in your group has several audio records.

12. Each time you record a new sound, record the information on your data sheet. A student who is not recording should fill out the data sheet.

13. Review and organize your field recording at the end of each day. The goal is to record as many vocalizations from as many animals as possible. These recordings may be used in Exercises 2 and 3.

Exercise 2: Sound Analysis Using Audacity

You will use Audacity software to analyze several audio recordings of mammalian sounds. Your instructor will provide you with access to the sound files you will analyze as part of this exercise. Additional sounds can be found at the Macaulay Library at the Cornell Lab of Ornithology (http://macaulaylibrary.org/browse/scientific /10520807) or at the Guide to Animal Sounds on the Net-Mammals (http://thryomanes.tripod.com/Mammal Sounds.html).

1. Download a lion growl .wav file from the Animal Sounds on the Net web page, and then launch Audacity.

2. Click on File > Open, and locate the audio file called "Lion growls" on your computer (Figure 15.6).

3. The waveform view is the default view in Audacity. It shows the amplitude (loudness) of the sound wave on the vertical axis and time on the horizontal axis. Each of the "notes" shows up as a separate set of peaks. Play the "Lion growls" audio file by clicking on the green Play button at the top left of the Audacity screen. A vertical line traces across the waveform as the sound file is played.

4. To the left of the waveform is an information box that lists the name of the file at the top ("Lion growl") and details about the recording below (Mono, 11025 Hz 32-bit, etc.). Click on the small black triangle next to the name of the audio file. A drop-down menu appears; click on Spectrogram from this list. Audacity now displays the

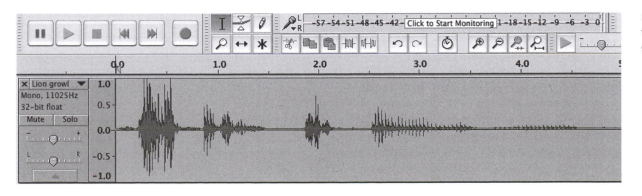

Figure 15.6 A waveform of a lion growl in Audacity.

Figure 15.7

A close-up of the last second of a lion growl in Audacity.

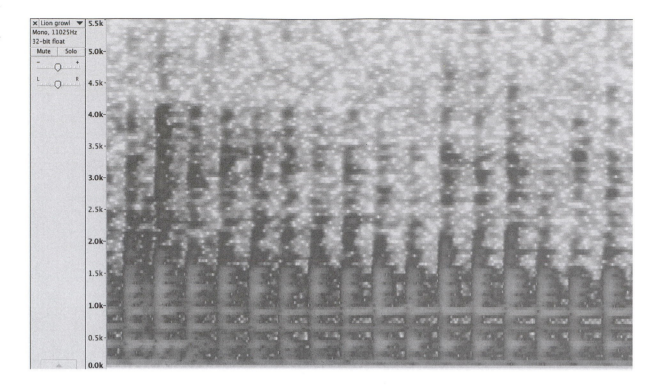

same file as a sonogram with the frequency (pitch) of the sound on the vertical axis in kilohertz and time in seconds on the horizontal axis. Loudness is indicated by colors.

5. Click Play to listen to this sound again. You should see the pitch (frequency) rising and falling during some of the notes. Focus on the small segment at the end of the sonogram where the lion is producing a deep-throated "purring" sound.

6. Drag the mouse over the section from roughly 9.5 seconds to 10.5 seconds. Click on View > Zoom to Selection. Then click on View > Fit Vertically (Figure 15.7). Click Play again to listen to this set of notes.

7. In the menu bar at the top of the page, click on Analyze and select Plot Spectrum. The Frequency Analysis window shows the "spectrum" of the sound—a plot of the amplitude (loudness) of the sound wave on the vertical axis and frequency on the horizontal axis. What frequencies are the loudest? Quietest? How do you think this would compare with other large cats? To find out, download a leopard growl from Mammals Sounds on the Net. Click on File > Open in the top menu bar and open the "Leopard growl" audio file.

8. Play the leopard file and select a one-second portion in the growl phase (e.g., from 1.0 to 2.0 seconds). Click on View > Zoom to Selection and then View > Fit Vertically. Drag the two Audacity windows so they are lined up. Do they appear similar? How would you determine which animal produces the lowest-frequency sounds?

9. Click in the "Lion growl" screen to make it active. Then click on Analyze > Plot Spectrum in the menu bar. You should see the power spectrum displayed for the "Lion growl" file. Repeat for the "Leopard growl" file to plot its power spectrum. The spectrum shows the highest peak at 905 Hz for the lion and at about 197 Hz for the leopard. At the bottom of the Frequency Analysis window for the leopard, the pitch is identified as F#3—that is, F-sharp in octave 3. What is the pitch of the highest peak for the lion?

10. Download and open the "Tiger file." Create a power spectrum of the "Tiger" file. How does it compare with a comparable section of the "Lion growl" file? (You will need to locate and re-analyze the lion file, focusing on the segment between 7.6 and 7.9 seconds.)

11. Next you will look at a very different type of sound. In Audacity, open the file named "*Myotis daubentonii*" provided on the website that accompanies this text (jhupbooks.press.jhu.edu/content/mammalogy-techniques-lab-manual) (Figure 15.8). Convert it to a spectrogram. How long is this sound clip? It appears that there are elements of the sound file that are greater than the 20 kHz displayed by default.

12. Click Play to listen to the sound. What do you hear? Probably nothing. Why is that? Can we use Audacity to generate an audible sequence for us?

13. Place the cursor over the *y*-axis, and a small magnifying glass icon appears. Click on the "20 k" near the top of the *y*-axis. Notice that there are indeed sounds above 20 kHz. Click again on the upper end of the *y*-axis until you reach about 100 kHz. You want to display the entire sound on the screen at one time. Hold down the Control key on a Mac (the magnifying glass has a minus symbol inside), and click just below the "100 k" on the *y*-axis. This shrinks the *y*-axis to fit inside the screen.

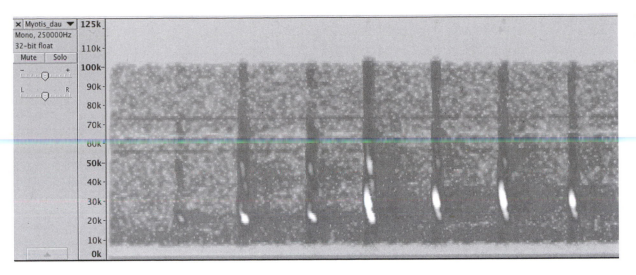

Figure 15.8
A sonogram of
Myotis daubentonii
showing the full
spectrum of
sound.

Stop when the *y*-axis runs from 0 k to 100 k (Figure 15.8). Now you can see the entire range of sound.

14. Play the sound. Can you hear it?

15. Click on Effect > Change Speed, and when the pop-up screen appears, set the slider to -60.0 (this slows it down by 60%). Click OK. Now Play the sound. You should now be able to hear the high-frequency echolocation signals.

16. Plot the power spectrum for this call. What is the loudest frequency? What is the lowest frequency? Is the lowest frequency within the human hearing range? Why don't we hear it? (Hint: look at the dB.) Now that you have seen how to analyze sounds using Audacity, search for a set of three to four mammal sounds on the web.

17. Create a set of hypotheses and predictions about these sounds. It may help to choose sounds from related species. For example, you might choose to compare large felids (lion and tiger) with smaller cats (bobcat or house cat), or you might compare the sounds made by various species of whales or primates. Alternatively, you could analyze several different calls from the same species (e.g., territorial calls versus alarm calls).

18. Download the .wav or .mp3 files to your computer from the sites listed previously, and analyze them in Audacity.

19. Write a report, in a format specified by your instructor, that describes your hypotheses and predictions and the results of your analysis.

Exercise 3: Playback Experiments Using Alarm Calls

In this exercise, you will conduct a playback experiment on squirrels, chipmunks, or some other locally abundant species (see Greene and Meagher, 1998, for a similar study). Your instructor will provide you with a set of audio tracks for playback in the field. You will record the squirrel's behavior in response to each playback. A variety of calls can be used. For example, your instructor may give you

- an alarm call from the squirrel species you are observing (but not from the same individual),
- the call of an aerial predator (bird of prey) native to your region (e.g., red-tailed hawk), or
- a segment of "white noise" that serves as the control sound.

Your field site should be a location where squirrels are abundant, easily visible, and relatively tolerant of human activity. It may help to set up feeding stations baited daily with sunflower seeds for several weeks in advance of the study to attract squirrels to the area you have chosen.

One trial consists of playing back one track. For each trial, you will make note of

- the squirrel's travel time to the feeding station,
- whether travel was direct, with no stops on the way to the feeding station, or periodic, with short stops to scan for intruders,
- handling time at the feeding station (total time spent handling, eating, or manipulating seeds),
- frequency and duration of vigilant behaviors (looking up, scanning area, etc.),
- residence time (total time at station),
- distance a subject animal ran to cover (after hearing the playback stimulus),
- recovery time (the time it took to re-emerge and resume activity), and
- number of seeds removed (count remaining seeds at the station after the trial).

The steps are as follows:

1. Locate the site where feeding stations have been in use, and set up your observation post about 10 meters away.

2. Set up the playback speaker (self-powered speakers that run on batteries) 3-4 meters from the feeding station, and camouflage it as best you can.

3. Add exactly 60 sunflower seeds to the feeding station (remove any remaining seeds from previous feedings).

4. Return to your observation post, and allow time for the subject to resume normal activity.

5. Begin trial #1 by making note of the pre-playback behavior of your subject animal for several minutes.

6. After your subject has been at the feeding station for 15–20 seconds, broadcast the first playback track (either a control sound or alarm call).

7. Simultaneously note the animal's behaviors during the playback (videotaping may be helpful).

8. After the playback, continue making note of behavior every 30 seconds for the next five minutes (i.e., the post-playback segment). This is the end of trial #1. Wait 30 minutes before proceeding to trial #2.

9. Count the seeds remaining at the feeding station, remove them, and place 60 new seeds at the station to begin trial #2.

10. Wait patiently for the subject to return to the feeding station and resume activity.

11. Play the second track while making note of the subject's response as before. This ends trial #2. Wait 30 minutes before proceeding to trial #3.

12. Count the seeds remaining at the feeding station, remove them, and place 60 new seeds at the station to begin trial #3.

13. Play the third track while making note of the animal's behavior as before. Thus, each subject receives all three playback treatments on the same day.

14. Collect your equipment and move to another study site (if instructed to do so), and repeat the process using a different subject animal.

After the experiment, the instructor will tell each group the type of call on each track. Because the observers do not know ahead of time what call is given at any time, this is called a blind experiment. Label each set of pre-playback, during playback, and post-playback behaviors with the track number and call type for your group. Groups will then pool their data for analysis. Your instructor will describe how the data should be graphed, analyzed statistically, and reported.

16 Quantifying Mammalian Behavior

Time Required
- One 3- to 4-hour lab period. Additional time may be needed.

Learning Objectives
- Understand how to categorize behaviors.
- Learn how to construct an ethogram.
- Learn different types of sampling in mammalian behavior.
- Practice behavioral observation techniques in the field.
- Understand how to avoid common problems.

Equipment Required
- Binoculars
- Field notebook
- Data sheets and clipboard
- Digital stopwatch

Background

By now you are familiar with the scientific method and hypothesis testing, where observations lead to questions, which in turn lead to specific hypotheses. Hypotheses are nothing more than a reformulation of a question into a formal, testable statement. For example, suppose you observed a group of foraging elk and wondered why certain individuals seem to spend more time looking around (i.e., vigilance behavior) than other members of the herd. This is essentially your question: why do certain individuals spend more time being vigilant than others? To create a testable hypothesis, you need to restate this question more formally, turning it into a statement that gives one potential explanation for the observation. You might state your hypothesis as "dominant individuals spend more time being vigilant than subordinate elk." This hypothesis is testable because you can collect behavioral observations on dominance level and amount of time spent being vigilant for each individual. In essence your hypothesis gives you a plan for collecting additional data. At this point, it is back to the field to collect the data needed to test your hypothesis. If your new data falsify your original hypothesis, then you choose an alternative hypothesis that potentially explains your observations. For example, you might restate your hypothesis as "individuals foraging on the perimeter of the herd spend more time being vigilant than those at the center of the herd."

This process sounds simple, but in practice you need to know enough about the behavior of elk to distinguish dominant behaviors from submissive behaviors and vigilance from other types of behavior. The first step in studying the behavior of a species is to construct an ethogram. In its simplest form, an ethogram is a catalog of an animal's behaviors (Altmann, 1974). Constructing ethograms requires careful observations of and notes on each behavior. The result is a complete list of behaviors with a definition of each. Ideally, each behavior should be unambiguous; it should not be possible to assign a new observation to more than one category in your ethogram. The level of detail in your ethogram depends on the goals of the study and the hypothesis being tested. For example, suppose you observe a raccoon digging in the muck at the margin of a stream, extracting a clam, breaking open the clam's shell on a rock, and extracting and eating the meat. Each of these events might be a separate category (with its own definition) in a study testing hypotheses of foraging behavior. However, if the goal is to construct a time budget for raccoon daily activity, you might combine all of these behaviors into one category called "foraging."

Avoiding Common Problems

Before you begin studying animal behavior, you should understand some common pitfalls of behavioral studies. Among the most important decisions is how many individuals to observe (i.e., to have a large enough sample size for statistical tests). It may be much easier to observe many behaviors in one individual, but this practice leads to pseudo-replication (treating the data as independent observations when they are in fact interdependent; Hurlbert, 1984). Observations need to be independent events. If you measure the same behavior multiple times in the same individual, the behaviors are not independent events (i.e., not true replicates). At times, your study design may require you to take repeated measures on the same animal (e.g., behaviors before and after an environmental disturbance). In general, though, the best way to characterize an animal's natural behavior is to watch as many different individuals as possible and then summarize their "average" behavior. Ideally, you can tell individuals apart. When that is not possible, you should watch a group large enough to create a low probability of watching the same individual twice.

Another problem that can crop up is observer bias. An observer who knows the hypothesis may unintentionally interpret behaviors as favoring the hypothesis (or ignore results that do not favor the hypothesis). Such bias can be avoided by conducting a "blind" study in which the observer does not know the hypothesis or its predictions in advance. Of course, this is not always practical, and observers should always strive for complete, honest, and unbiased records of behavior.

Often, two or more observers will record data to be pooled for analysis in order to, for example, yield larger sample sizes. In such cases, it is essential to know that each observer is recording the same thing. For example, one observer might score a behavior differently than another observer. One way to prevent inter-observer variability is to properly train the observers beforehand and cross-check their results. Observers should also be randomly assigned to study groups. This is especially important if there are separate treatment and control groups. Having one observer record only treatment groups and another observe only control groups leads to invalid results. Behavioral studies present other pitfalls as well, but these are the most common.

Exercise 1: Building an Ethogram

In this exercise, you will construct an ethogram for a species chosen by your instructor (squirrels are an obvious choice in many regions, because many species are diurnal and easily observed). In your field work, you first must characterize the range of behaviors you see. Define behaviors in terms of the animals' actions, not by function. For example, if you see a squirrel bury an acorn, you would record it as "burying" and define it as "digging and placing an acorn in the ground, and covering it." You would not record it as "storing food" because this implies a function or interpretation you did not observe directly—that the animal would return at some later point to retrieve the acorn. One of the hardest lessons to learn when studying animal behavior is to avoid projection (of either adaptive function or anthropomorphism).

1. Find a suitable spot to observe your target animals. Make sure you are far enough away to avoid disturbing the animals' behavior but close enough to see what is happening.

2. Sit down, get comfortable, and begin to observe and record behaviors. Avoid excess movements or noise.

3. In your field notebook, record all the different kinds of behavior you observe. Only after you have built a catalog of possible behaviors for your animals will you be ready to quantify those behaviors in terms of frequency, duration, and context.

4. Initially, you should note the context in which behaviors are performed. "Context" includes location (habitat type, proximity of major features of the habitat), other organisms (same or different species), physical conditions (time/temperature/light), previous behavior, and individual characteristics (male/female, nutritional state, reproductive state, size) if known. One way to do this is to create a log of all the behaviors as they occur, identifying the next behavior in the sequence and, if possible, noting how long each behavior lasted.

Now that you have a reasonably complete set of field observations, it is time to begin organizing the behavioral catalog or ethogram. The ethogram is a list of behaviors and an explicit definition for each. The catalog allows you to organize your data sheets so that you can quantify behavior. For example, some grey squirrel (*Sciurus carolinensis*) behaviors are shown in Table 16.1.

Once you have the basic catalog with objective definitions, you can organize your catalog according to behavior types. In Table 16.1, "running," "climbing up," and "climbing down" might fall within a category called "movement," whereas "chirp" and "chatter" might fall under "vocalization." Such a hierarchy will let you organize your data sheets and your analyses more easily. You should also assign each behavior category a unique, easy-to-recall letter code (e.g., R for running and D for climbing down).

Table 16.1 Sample behavior catalog for a grey squirrel

Behavior	Definition
Running	Bounding movement over the ground
Climbing up	Vertical movement away from the ground
Climbing down	Vertical movement toward the ground
Chasing	Running or climbing behind a conspecific

Exercise 2: Sampling Behaviors

Now that you have completed your catalog of behaviors, you are ready to quantify how your animal spends its time. You are expected to observe your study animal for at least three hours. You should aim to observe at least four individuals in this time period, devoting an approximately equal amount of time to each individual you observe. Be careful about when you observe them. It is best if you can observe each individual multiple times (i.e, three 20-minute observations of the same individual is better than a single 1-hour observation of that individual). Distribute your observation periods through the day so that you can get a generally representative sample of behavioral patterns. Alternatively, restrict your observations to particular times of the day and distribute your observations evenly within those constraints.

There are a number of methods for sampling behaviors:

- What you have done so far is called **ad lib sampling**. In ad lib sampling, behaviors are recorded in the order they occur. Bias can creep into ad lib sampling if the observer tends to notice and record certain behaviors over others.
- In **scan sampling**, the observer scans the entire group of animals and records what each animal is doing. This approach works well for smaller groups. It is used to construct time budgets and/or for studying individual variation. A further advantage is that it is unbiased because every animal is sampled each time. Typically, scans are taken every 30 seconds, every minute, or at some other predetermined interval.

- **Focal animal sampling** involves selecting and following a single individual. Its behaviors are recorded for a standard time (or until the animal moves out of sight). If a focal individual moves out of your view, then you start a new sequence of observations on a new focal individual. Selecting the focal animal can be systematic (e.g., follow only young animals) or randomized (select a random number from a table, then follow the nth individual encountered). The main disadvantage of focal animal sampling is that you can only record data from one animal at a time, so it takes longer to acquire an appropriate sample size.

Organized data sheets are critical for behavioral studies, because they allow you to record data quickly and efficiently during the observation period and to tabulate the results accurately afterward. A well-written catalog of behaviors will allow you to create a useful data sheet, which in turn simplifies the quantification phase of the study. A simple example data sheet is shown in Table 16.2.

1. Create a data sheet from your ethogram.
2. Return to the field and perform a scan sample every one minute for one hour.
3. Fill in your data sheet by recording how many individuals are exhibiting each behavior at the time of the scan.
4. Perform a focal animal sample by arbitrarily choosing one animal in the group and recording its behaviors on a new data sheet. Record each behavior in turn and the amount of time that behavior lasts.

Table 16.2 Sample data sheet for a study of aggressive behaviors

Time (1-minute intervals)	Behavior (code)					
	Chase	Fight	Attack	Bite	Submit	Retreat
14:00						
14:01						
14:02						

Exercise 3: Creating a Time Budget

In Exercises 1 and 2 you learned how to catalog and sample behaviors. The more interesting questions in animal behavior require us to detect relationships or patterns among the various behaviors animals perform. One way to detect patterns is to record the time and sequence of each behavior. The objective is to determine the probability that one behavior will be followed by (or preceded by) another behavior. Such sequence diagrams and time budgets can be derived from focal studies. In this exercise, you will create a time budget analysis.

A time budget indicates the amount or proportion of time that animals spend in different behaviors or in performing different classes of behavior. To construct a time budget, you must add the time spent on each behavior for each subject animal. Then, out of the total observation time, calculate the proportion of time spent on particular behaviors. When you are able to observe several different individuals, then each individual can be a "replicate" or independent estimate of time allocation of the behavior of interest. You summarize your data by doing a time budget for each individual, then calculating mean values for each behavioral measure of interest. An example appears in Table 16.3.

1. Calculate the mean time spent for each behavior for all animals in the study, and convert this figure to a percentage of total observation time.

2. Create a table similar to Table 16.3 to summarize your results, and write up a short results section in journal format.

3. List at least one hypothesis that you could test using the data from your time budget.

Table 16.3 A partial time budget for large mammalian herbivores on the National Bison Range, Montana

Species	Inactive behaviors (%)		
	Lying down	Standing	Ruminating
Antilocapra americana	33±15	14±7	14±3
Ovis canadensis	38±20	6±6	15±15
Odocoileus virginianus	50±60	5±6	11±4
Odocoileus hemionus	30±27	11±14	12±13
Cervus elaphus	48±33	7±9	13±20
Bison bison	40±31	14±5	20±3

Note: Data are from Belovsky and Slade (1986).

Exercise 4: Creating a Transition Diagram

The second method for analyzing behavior patterns is to derive a transition matrix. A transition matrix expresses the probability that one behavior is followed by another. For example, we might expect that digging a hole has a 0.85 probability (85% of the time) of being followed by burying an acorn, but digging a hole has only a 0.15 probability of being followed by chasing another squirrel. By linking "chains" of behaviors, a transition matrix is a technique for looking at relationships between different behaviors. Creating a transition matrix requires knowing the sequence in which each behavior was performed.

In a transition matrix, behaviors are listed in rows and columns in a table. The cells of the table show the relative frequency that a reference (row) behavior is followed by the next (column) behavior. A simple example of a transition matrix is shown in Table 16.4.

Suppose your first sequence of behaviors is a chase followed by an attack, followed by a bite. You would

Table 16.4 An example of the raw data for aggressive behaviors used to create a transition matrix

Beginning behavior	Followed by behavior (number of occurrences)						
	Chase	Fight	Attack	Bite	Submit	Retreat	Total
Chase	0	21	8	3	67	54	153
Fight	31	0	2	45	56	87	221
Attack	23	69	0	15	45	91	243
Bite	1	21	0	0	78	23	123
Submit	0	1	6	0	0	0	7
Retreat	0	0	0	0	5	0	5
Total							752

Table 16.5 Transition percentages for the aggressive behavior data presented in Table 16.4

Beginning behavior	Followed by behavior						
	Chase	Fight	Attack	Bite	Submit	Retreat	Totals
Chase	0%	14%	5%	12%	44%	35%	100%
Fight	14%	0%	1%	20%	25%	39%	100%
Attack	9%	28%	0%	6%	19%	37%	100%
Bite	1%	17%	0%	0%	63%	19%	100%
Submit	0%	14%	86%	0%	0%	0%	100%
Retreat	0%	0%	0%	0%	100%	0%	100%
Totals							100%

score the chase-attack sequence as a 1 in row 1 (chase), column 3 (attack) and a 1 in row 3 (attack), column 4 (bite). You continue tallying each sequence until all the aggressive behaviors have been scored and added to your table. The result is a matrix that shows the number of times each behavior listed in column 1 on the left ("Beginning behavior") is followed by each behavior listed in the row across the top ("Followed by behavior"). Thus, it gives the number of times there was a transition from one type of behavior to another (a behavior can follow itself if two of the same behaviors occurred in sequence).

You next calculate the percentage of times that a particular action pattern follows another (given the first action pattern). In Table 16.4, for example, 21 chases were followed by fights out of a total of 153 chases observed. The transition probability is $21/153 = 0.137 = 13.7\%$. Table 16.5 illustrates the transition percentages from the raw data in Table 16.4.

Please note that transition percentages across a row must sum to approximately 100, but individual columns do not need to add to 100%. Can you explain why this would be the case?

A transition diagram (or kinematic diagram) is simply a flowchart of the behaviors. The steps to make one are as follows:

1. Start by drawing a circle or box, and place the abbreviations for one behavior in the box. You can begin with any behavior. You may want to add sample sizes in parentheses (from the raw data matrix, right-column totals) into the box with each behavior (Figure 16.1).

- Draw arrows to the other behavior circles that followed this behavior (Figure 16.1).
- Next to each arrow place the percentage from the transition matrix. Be sure to note the direction of the arrow.

Transition diagrams allow you to see quickly which behaviors follow which others and give the frequency

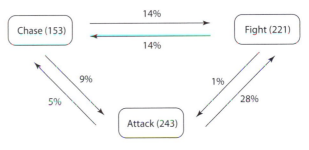

Figure 16.1 Preliminary transition diagram for the data in Table 16.5.

with which these sequences of behavior occur. By examining transitions in different individuals or in different conditions, you may learn about "complexes" of coordinated behaviors. Also, you can ask questions about how different conditions influence the flow of behaviors through time.

Exercise 5: Creating a Dominance Hierarchy

A dominance hierarchy analysis is a record of winners and losers in encounters that is used to determine dominance ranks among individuals. For example, suppose you were interested in establishing the dominance hierarchy for a small herd of elephants from Amboseli National Park, Kenya. One such family group hierarchy is shown in Table 16.6 (from Archie et al., 2006).

In Table 16.6, the letters represent one individual in the CB family unit, and the intersection of rows (aggressor) and columns (loser) shows the number of interactions won by aggressor (rows). For example, ele-phant A won 6 times against elephant B and 7 times against elephant D, but only once against elephant C. Elephant A is clearly the dominant animal in the group. Can you rank dominance of these elephants based on information in this table?

Dominance hierarchies often are linear, showing a specific pecking order within the group. In the elephant example in Table 16.6, elephant A dominates everyone else, elephant B dominates everyone but elephant A, and so on. In other cases, the hierarchy may not be quite so linear. One way to test for linearity is to use Landau's H.

$$H = \left(\frac{12}{n^3 - n} \right) \sum \left[v_a - \frac{n-1}{2} \right]^2$$

where n is the number of animals in the group and v_a is the number of animals dominated by animal a (i.e., the number of individuals that individual outranks). Landau's H ranges from 0 to 1, with values near 1 being more linear hierarchies and 0 indicating no hierarchy.

1. Calculate Landau's H for the elephant group in Table 16.6.

2. Construct a dominance hierarchy matrix from data you have collected, and determine whether your animals exhibit a dominance hierarchy.

Table 16.6 Dominance matrix for the CB family unit of elephants from Amboseli National Park, Kenya

	A	B	C	D	E	F
A		6	1	7	6	2
B			1	3	6	1
C				2	1	1
D	3				5	2
E		1				
F						

Note: Data are from Archie et al. (2006).

Exercise 6: Dominance Hierarchy Analysis

You will now use a Java applet (an online software tool written in the Java programing language) to examine dominance hierarchies in an example elephant data set and in data you collect on your subject species in the field.

1. Using your web browser, go to http://caspar.bgsu .edu/~software/Java_Support/1Hierarchy.html. This software is part of a collection of freeware programming functions for analyzing behavioral data produced by Dr. Robert Huber (available on the internet at http:// iEthology.com/).

2. Assuming your browser supports Java, you should see a description of the Java applet followed by a series of three contingency tables at the bottom of the page (Figure 16.2).

3. You can set the size of the data matrix using the drop-down buttons at the very bottom of the web page. In your elephant example from Table 16.6, we have a 6×6 matrix. Set the applet's matrix size to 6×6, using this button at the bottom of the page.

4. Click on the first available cell (e.g., cell 1B) in the applet matrix; it turns light blue when it is available for data entry. Use the data from Table 16.6 to fill in the upper-right part of the applet table (Figure 16.3). You must click on each cell, type in a number, and then click on the next cell. If you make a mistake, you have to reload the page (there is no reset button to clear the table).

5. With the data entered, you can now examine the output of several measures. First, click on the Linearity button on the left side of the analysis table (Figure 16.4). You will see your value for Landau's H listed here. Does this value correspond well with the value you calculated by hand? Does this value suggest that the dominance hierarchy for this family unit is linear?

6. Next click on the Ranks button in the analysis table. This will display the calculations for several ranking methods (Figure 16.5). The first two measures are the ordinal ranks and dominance index (from Theraulaz et al., 1992). These both indicate that elephant A in our example is dominant to all others in the group, and that elephant B is dominant to all others except elephant A. The third column shows the dominance activity index (DAI) calculated as follows (from Bartos, 1986):

$$DAI = \log \left\{ \frac{(P+0.1)^2}{N+0.1} \right\} + 1$$

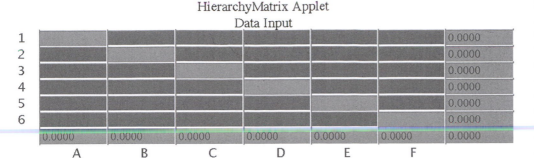

HierarchyMatrix Applet
Data Input

	A	B	C	D	E	F	
1							0.0000
2							0.0000
3							0.0000
4							0.0000
5							0.0000
6							0.0000
	0.0000	0.0000	0.0000	0.0000	0.0000	0.0000	0.0000

Analysis

TauKr	
WhoFights	
WithWhom	
Linearity	
Ranks	

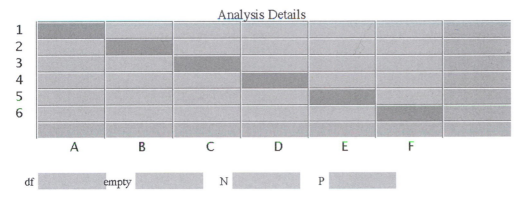

Analysis Details

	A	B	C	D	E	F
1						
2						
3						
4						
5						
6						

df _____ empty _____ N _____ P _____

Figure 16.2 The Java applet for the dominance hierarchy analysis. *The analysis was performed using a collection of freeware programming functions available at http://iEthology.com/.*

The final two columns display ranks based on the Boyd and Silk (1983) method and the Batchelder-Bershad-Simpson (BBS) method (Jameson et al., 1999). The Boyd and Silk method will not work properly if we lack information about interactions between some pairs of individuals or if the group includes individuals who are never dominated by others. The BBS method, however, does provide stable ranks even if both of these conditions are not met. In the BBS method, an animal's rank depends on (1) its proportion of wins, (2) its proportion of losses, and (3) the scores of the others it encountered (Jameson et al., 1999). In this case, low scores represent higher ranks. The BBS model uses the following equation to provide initial estimates of rank:

$$s(a_i) = \left[\frac{a(2W_i - N_i)}{2N_i} \right]$$

where a is a constant ($a = \sqrt{2\pi} = 2.50663$), W_i is the number of encounters in which animal a_i won, and N_i is the total number of encounters for animal a_i.

Each animal is assigned an initial score $s(a_i)$ based on the proportion of interactions it won. Then, a second equation is used to rescale (recursively) all animals re-

HierarchyMatrix Applet
Data Input

	A	B	C	D	E	F	
1		6.0000	1.0000	7.0000	6.0000	2.0000	22.0000
2			1.0000	3.0000	6.0000	1.0000	11.0000
3				2.0000	1.0000	1.0000	4.0000
4	3.0000				5.0000	2.0000	10.0000
5		1.0000					1.0000
6							0.0000
	3.0000	7.0000	2.0000	12.0000	18.0000	6.0000	48.0000

Figure 16.3 Example of the dominance data entered into the applet. *The analysis was performed using a collection of freeware programming functions available at http://iEthology.com/.*

peatedly until their scores do not change. The second equation is

$$s(a_i) = \left[\frac{2(W_i - L_i)}{N_i} \right] + Q_i$$

where L_i is the number of encounters animal a_i lost, and Q_i is the mean score of animals that a_i encountered.

Use the data you gathered from your field observations to rerun the dominance hierarchy applet. Write a short summary of your results.

Figure 16.4 Linearity analysis for the elephant data. *The analysis was performed using a collection of freeware programming functions available at http://iEthology.com/.*

TauKr	
WhoFights	
WithWhom	
Linearity	Landau h: 0.9714; circ. triads: 0; Appleby K: 1.0000
Ranks	

Analysis Details

	A	B	C	D	E	F
1		1.0000	1.0000	1.0000	1.0000	1.0000	5.0000
2	0.0000		1.0000	1.0000	1.0000	1.0000	4.0000
3	0.0000	0.0000		1.0000	1.0000	1.0000	3.0000
4	0.0000	0.0000	0.0000		1.0000	1.0000	2.0000
5	0.0000	0.0000	0.0000	0.0000		0.5000	0.5000
6	0.0000	0.0000	0.0000	0.0000	0.5000		0.5000
	0.0000	1.0000	2.0000	3.0000	4.5000	4.5000	15.0000

df 20   empty 14   N 30   P

Figure 16.5 Analysis of the ranks of the elephant family unit using five methods. *The analysis was performed using a collection of freeware programming functions available at http://iEthology.com/.*

Analysis

TauKr	
WhoFights	
WithWhom	
Linearity	
Ranks	Boyd&Silk: -1 iterations; BBS: -1 iterations

Analysis Details

	A	B	C	D	E	F
1	1.0000	0.8696	5.8700	❓	-1.1030		
2	2.0000	0.5882	3.6650	❓	-0.1565		
3	3.0000	0.6000	2.5209	❓	-0.3129		
4	4.0000	0.4000	2.6905	❓	0.1449		
5	5.0000	0.0526	-1.7053	-201.8165	1.4051		
6							
	ordinal	dom. index	DAI	Boyd&Silk	BBS		

df 12   empty 8   N 20   P

Exercise 7: Social Network Analysis Using igraph

Social network analysis (SNA) is a method for investigating social relationships. It uses the mathematics of both network and graph theory (Coleing, 2009; Wittemyer et al., 2007). In simplest terms, it creates a network of nodes (individual animals, people, or things) and a set of edges or links between those nodes. For example, the edges are the relationships or interactions connecting individuals in the network. SNA is used in sociology to visualize friendships or business relationships. In epidemiology it is used to visualize disease transmission within a population. In national security it is used to determine leadership in terrorist networks. In animal behavior it is used to understand social structure and dominance relationships, among many other applications (Farine and Whitehead, 2015; Norscia and Palagi, 2015). These networks are visualized through graphs in which nodes are represented as circles and ties are represented as lines between circles.

In this exercise, you will explore a set of social relationships among a hypothetical network of lemurs where some individuals are associated more strongly with one another than others. This could be because these individuals are related, because they forage together more frequently, or simply because they tend to groom one another more often. The first step is to create an adjacency matrix in Excel for grooming behavior.

Table 16.7 Grooming data for 13 lemurs from Madagascar

	TT	BI	BV	CA	CO	CS	CT	CV	MY	NI	PD	PG	SA
TT	0	0	0	8	12	0	0	0	0	2	0	11	0
BI	0	0	1	4	3	0	1	0	0	2	0	3	0
BV	6	0	0	17	9	0	1	0	0	0	6	6	0
CA	0	0	0	0	0	0	0	0	0	0	0	0	0
CO	0	0	0	4	0	0	0	0	0	2	0	1	0
CS	1	0	1	1	1	0	2	0	0	0	0	3	1
CT	0	0	1	4	2	0	0	0	0	2	3	2	0
CV	0	0	1	1	1	0	1	0	0	1	5	0	0
MY	1	3	0	0	1	3	1	1	0	0	4	4	2
NI	0	0	0	1	0	0	0	0	0	0	0	0	0
PD	0	0	0	1	0	0	1	0	0	1	0	1	0
PG	1	0	0	2	4	0	0	0	0	2	0	0	0
SA	1	6	3	3	2	1	6	2	0	1	15	2	0

Note: Lemur names are given as two-letter abbreviations, and the numbers represent the number of grooming events between each lemur pair.

Table 16.8 A table of the sex of each of the 13 lemurs in Table 16.7

ID	Sex
TT	F
BI	F
BV	F
CA	F
CO	F
CS	F
CT	F
CV	M
MY	M
NI	M
PD	M
PG	M
SA	M

Open Excel and create the file shown in Table 16.7. It depicts the data for 13 lemurs from Madagascar. Here the names are given as two letters across the top row and in the first column. The numbers represent how often the pair groomed each other. For example, lemur TT and lemur CA groomed each other 8 times, but TT never groomed lemur BV. Save this file as "Lemur_adjacency.csv". Excel's default is to save it as an .xls file, but be sure to save it as a .csv file.

Next, create a second Excel matrix with information on the sex of each lemur, like that shown in Table 16.8. Save this file as "Lemur_attributes.csv".

Open RStudio and load the igraph package (Csardi and Nepusz, 2006).

```
install.packages("igraph")
library(igraph)
require(igraph)
```

Import the lemur adjacency data and call it "dat."

```
dat=read.csv(file.choose(), header=TRUE,
row.names=1, check.names=FALSE)
```

This command asks you to locate the .csv file you want to load and lets R know that the first line is a header (not data). With the "Lemur_adjacency.csv" file loaded, you need to tell R to make it a matrix. Type the following to create a matrix called m:

```
m=as.matrix(dat)
```

Now you tell R that these data are weighted, which means that some individuals not only interact with others, but those interactions are given a weight (higher numbers indicate stronger or more frequent interactions):

```
net=graph.adjacency(m, mode="undirected",
weighted=TRUE, diag=FALSE)
```

In social network analysis, a node or vertex is a circle representing an individual lemur and an edge is a line connecting two lemurs. When weighted=TRUE, each edge has a value assigned to it. To see the edge values, enter

```
E(net)$weight
```

Now you can graph the network of lemur interactions. Enter the following code all on one line:

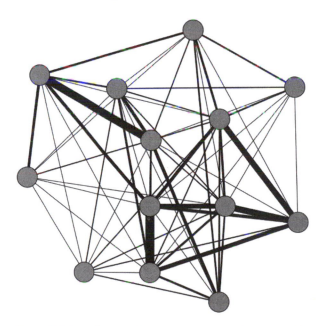

Figure 16.6 A social network analysis diagram produced in igraph R package. Each lemur is represented by a circular node, and the lines connecting the lemurs represent interactions between individuals.

```
plot.igraph(net, vertex.label=NA,
layout=layout.fruchterman.reingold, edge.
color="black", edge.width=E(net)$weight)
```
You should now have a plot like the one in Figure 16.6. Here the nodes are the lemurs, and the interactions are coded as the thickness of the line between a pair of lemurs.

Suppose you want to know which lemurs are males and which are females. You can add this to your network plot by calling on the "Lemur_attributes.csv" file. Enter the commands below to color the nodes yellow for females and blue for males:
```
a=read.csv(file.choose())
V(net)$Sex=as.character(a$Sex[match(V(net)
$name, a$ID)])

V(net)$color=V(net)$Sex
V(net)$color=gsub("F", "yellow",
V(net)$color)
V(net)$color=gsub("M", "light blue",
V(net)$color)
```
To plot the new network, enter the following all on one line:
```
plot.igraph(net, vertex.label=NA,
layout=layout.fruchterman.reingold, edge.
color="black", edge.width=E(net)$weight)
```
The plot is now color-coded by sex (Figure 16.7). However, Figure 16.7 does not identify the individual lemurs. To add the names (labels) to the graph, use the following command:
```
V(net)$name
```
This command calls the names for the nodes in graph "net" and prints them in the Console window as
```
[1] "TT" "BI" "BV" "CA" "CO" "CS" "CT" "CV"
"MY" "NI" "PD" "PG" "SA"
```

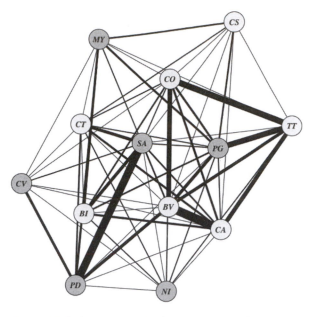

Figure 16.8 The lemur data coded for sex and name of the individual.

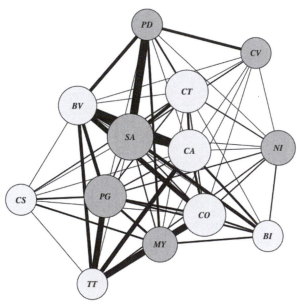

Figure 16.9 The lemur data re-coded to show the magnitude of conspecific interactions as the size of the individual's node.

You can now add the names to the graph by typing the following all on one line:
```
plot.igraph(net, layout=layout.fruchterman.
reingold, vertex.label=V(net)$name, vertex.
label.font=4, edge.color="black",
edge.width=E(net)$weight)
```
This command is similar to the one used above, but it now includes **vertex.label=V(net)$name and vertex.label.font=4** attributes. The first sets the names of the vertices to **V(net)$name** and the second makes the font a little larger. The plot should now look like Figure 16.8.

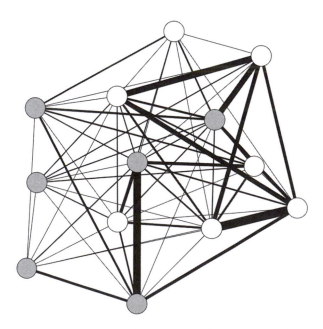

Figure 16.7 The same data from Figure 16.6 with lemurs coded by sex. White nodes are female lemurs, and grey nodes are males.

Next, change the size of the animal's node (vertex) to reflect the number of edges connected to it. To do this, enter the attribute **vertex.size=degree(net)*4**, into the command as shown:

```
plot.igraph (net, layout=layout.
fruchterman.reingold,
vertex.size=degree(net)*3,
vertex.label=V(net)$name,
vertex.label.font=4,edge.color="black",
edge.width=E(net)$weight)
```

The plot now looks like Figure 16.9, where the size of each animal's node (vertex) is proportional to the number of interactions. Play around with adding additional attributes to the plot by following the examples in the igraph documentation. To save a copy of your plot, simply click on the Export tab in the plot window.

Many questions can be explored using social network analysis. You might have a data set for aggressive encounters among a social group and want to explore the dominance hierarchy. Alternatively, you may want to look at how social networks change when one individual is removed (e.g., an alpha individual) or how the group structure changes in a fission-fusion society. To explore more about social network analysis, review some of the examples in the igraph R package at http://igraph.org/r/#docs.

17 Optimal Foraging Behavior

Time Required
- One 3- to 4-hour lab period for each of the exercises

Learning Objectives
- Understand the basics of optimal foraging theory.
- Understand how energy and profitability are calculated.
- Understand how handling time, search time, and other factors shape foraging decisions.
- Learn how animals make decisions about which patch to forage in.
- Apply optimal foraging theory to squirrel foraging in the field.

Equipment Required
- Clipboards
- Data sheets
- Peanuts (shelled and unshelled) and sunflower seeds
- Two-foot-square plywood trays, painted white
- Tape measures
- Stopwatches
- Binoculars

Background

All mammals are predators at some level. Wolves are predators that eat elk, deer, and moose. Kangaroo rats are capturing helpless seeds. Whether they are carnivores or herbivores, all mammals must search for and capture food. In doing so they must make choices about what type of food to search for, how long to spend searching, and where to search. Optimal foraging theory attempts to understand why certain foraging behaviors are adaptive (Krebs, 1978). The basic prediction is that animals will forage in a way that maximizes net energy intake per unit of time (Kolmes et al., 1997). Doing so ensures maximum fitness in terms of survival and overall reproductive success.

Net energy intake rate is determined by several factors, including

- energy spent searching for food,
- energy value of the prey item,
- energy spent handling or capturing prey,
- the abundance of the food items in the animal's environment, and
- the distribution of the food items in the environment.

Thus, there are costs and benefits associated with foraging (Kolmes et al., 1997; Charnov, 1976). Over time, natural selection shapes behaviors that maximize fitness. It is therefore assumed that an animal's foraging behaviors are an important component contributing to its overall fitness.

Each food item has its own handling time, denoted by h. It is the amount of time it takes the predator to subdue, prepare, and consume a prey item once it has located the item. In addition, each prey item has a unique caloric value (i.e., energy content) denoted by E. It is the amount of energy that can be extracted by one prey item via absorption across the intestines minus the energy spent to chew and digest the item. Thus, each prey item, whether it is a gazelle or a seed, has a unique profitability, denoted by P, which is the net caloric value of a prey item divided by the handling time (units are typically calories/minute or joules/second).

$$P = \frac{E}{h}$$

Virtually all mammals live in environments that contain a mixture of food items (i.e., prey species). Some prey are more profitable than others, and some prey are more abundant than others. Optimal foraging theory predicts that predators should select prey items in such a way that the prey profitability and prey density (abundance) combine to yield the greatest net energy gain. In other words, if the most energy-rich prey is rare, it may not pay to spend all the energy needed to search for a rare prey when other less profitable prey are easily at hand. Where multiple prey are available, what is the optimal foraging strategy for the predator?

From the predator's point of view, "prey density" really signifies "prey encounter rate." The denser the prey item, the more frequently the predator will encounter it.

Encounter rate is denoted by λ, where the encounter rate for a prey of type i is λ_i = total encounters with prey type i divided by total search time. For example, a foraging otter might encounter 20 fish in a 10-minute interval, yielding an encounter rate for fish of λ_{fish} = 2.0 fish per minute.

The encounter rates of each prey species are important because they influence search times. To see why, consider a predator searching for food in an environment where prey are scarce. Assume it takes the predator t seconds to locate a particular type of prey. If we double the number of prey types searched for, and we assume equal prey densities for both prey species, then there should be twice as many encounters in the same time period. Stated differently, $t/2$ search time is cut in half. If we now add a third prey type (of equal density), the encounter rate becomes $t/3$ and so on. Thus, search time is inversely proportional to the number of prey types available (Figure 17.1).

Notice that when there are few prey species, the graph is steep, but when more prey types are present it flattens out. In practice, this means that a small increase in the number of prey species can dramatically reduce search costs for the predator. Thus, a predator cannot be too fussy about which prey species it seeks, or it may end up spending huge amounts of time searching.

Of course, so far we have assumed that each prey species is equally abundant. This is rarely the case in the real world. If a predator is in a good habitat where prey of all species are abundant, then search times will be small for each prey type. However, if prey are less abundant in poor habitats, then search times increase proportionally (Figure 17.2).

Recall that predators should try to maximize net caloric intake rate (E/T). In this case, the model is concerned with maximizing energy, but it is possible to use

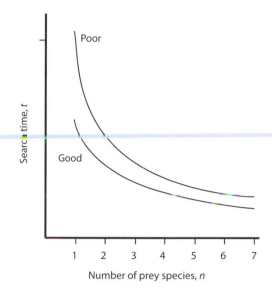

Figure 17.2 Search time (t) and the number of prey types (n) in a good and poor habitat. Notice that the steepness of the line for small numbers of prey (n) increases more sharply in a poor habitat.

other "currencies." For example, the energy-maximizing model ignores the forager's need to avoid predators too. We might want a model that minimizes the time spent foraging (and therefore the risk of being preyed upon).

In a simple model of prey choice, a predator specializes on only one prey type. Thus, its net caloric intake rate would be

$$\frac{E}{T} = \frac{\lambda_1 E_1}{1 + \lambda_1 h_1}$$

where λ_1 is the encounter rate of the prey, E_1 is the energy content in calories of the prey item, and h_1 is the handling time for that prey.

Suppose, however, we want to model how a forager would maximize its energy intake rate when given a choice of two different prey species. In this case, the encounter rates of the two prey are different (i.e., the prey have different densities), and we can denote them as λ_1 and λ_2. The net caloric values of the two prey species are denoted as E_1 and E_2, and their handling times are h_1 and h_2. To simplify, assume the predator is not selective (will not ignore one prey over the other). This means that in t minutes of searching, the predator will have encountered prey 1 at a rate of $\lambda_1 t$ and prey 2 at $\lambda_2 t$. The overall caloric intake rate is

$$\frac{E}{T} = \frac{t(\lambda_1 E_1 + \lambda_2 E_2)}{t(1 + \lambda_1 h_1 + \lambda_2 h_2)} = \frac{\lambda_1 E_1 + \lambda_2 E_2}{1 + \lambda_1 h_1 + \lambda_2 h_2}$$

Similarly, if there are n different prey species in the predator's diet, then nonselective predation would give a net caloric intake rate of

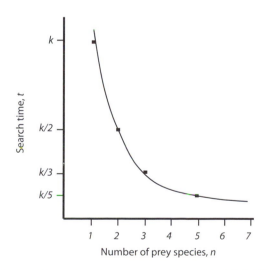

Figure 17.1 The inverse relationship between search time (t) and the number of prey types (n) in a predator's diet.

$$\frac{E}{T} = \frac{\lambda_1 E_1 + \lambda_2 E_2 + \cdots + \lambda_n E_n}{1 + \lambda_1 h_1 + \lambda_2 h_2 + \cdots + \lambda_n h_n}$$

The math looks complicated but is really quite simple (detailed examples and mathematical proofs can be found in Kolmes et al., 1997). These equations can be used to determine exactly which prey species should be included in the forager's diet. For example, if we know the profitability of prey species 1 is greater than that of prey species 2,

$$\frac{E_1}{h_1} > \frac{E_2}{h_2}$$

and we want to know if it is more profitable to add prey 2 to the diet or simply feed only on prey 1, then we need to know which diet maximizes net caloric intake rate. In this case prey species 2 should be added to the forager's diet only if its profitability is greater than the net caloric intake rate for prey species 1 alone—in other words, if

$$\frac{\lambda_1 E_1}{1 + \lambda_1 h_1} < \frac{E}{h}$$

Similarly, a third prey species should be added to a diet consisting of prey species 1 and 2 only if the intake rate of the first two prey species is less than for all three prey species. The equations above can be used to create a set of foraging rules. If there are n prey species in a possible diet and they are ranked by profitability, then

1. the forager should always include the most profitable prey in its diet, and
2. add additional prey species into the diet until the profitability of a new species is less than the net caloric intake rate of the diet without that prey species.

These rules allow you to test whether an animal is foraging optimally. For example, consider the data in Table 17.1.

What is the optimal diet for this hypothetical fox? Meadow voles have the highest profitability (50 kcal/min) and should be included in the fox's diet according to rule #1. We can calculate the net intake rate for a diet consisting only of voles as

$$\frac{E}{T} = \frac{0.2(160)}{1 + 0.2(3.2)} = 19.5$$

The profitability of deer mice (42 kcal/min) is greater than the net intake rate of the vole-only diet (19.5 kcal/min), so rule #2 says that deer mice should be included in the diet. Voles and mice together are an optimal diet providing roughly 31.1 kcal/min of net energy:

$$\frac{E}{T} = \frac{0.2(160) + 3.0(25)}{1 + 0.2(3.2) + 3.0(0.6)} = 31.1$$

The profitability of masked shrews (25 kcal/min) is less than the intake rate from a diet of both voles and mice (i.e., less than 31.1), so shrews should not be added to the fox's diet. If the hypothetical fox was observed eating all three small mammals, then the hypothesis that the fox was foraging optimally would be falsified.

One additional point to make is that if the encounter rate increases, the net caloric rate for any diet increases, yet the profitability of each prey item remains the same. Can you explain why? The prediction is that if encounter rates increase, the optimally foraging animal should become more selective and eliminate one or more prey species from its diet. For the hypothetical fox example, encounter rates might increase as winter gives way to spring and summer, and the diet breadth of the fox should be narrower in the summer than in the winter. This prediction can also be tested with data collected in the field.

Foraging in Patches

In the wild, encounter rates usually vary across the habitat, and the habitat is said to be patchy. Thus, a given environment may have a range of patches, from poor-quality patches (where prey are absent or scarce) to high-quality patches (where prey are diverse and abundant). Foragers move between patches looking for food. If patches vary in quality, then we can assign a profitability to each patch. This leads to a number of interesting questions:

- How many patches should a forager visit?
- How long should a forager spend searching within a patch?
- What is the optimum time to quit one patch for another?

Table 17.1 Hypothetical diet of a fox

Item	Kilocalories	Minutes	Kilocalories/minute	Encounters/minute
Meadow vole	160	3.2	50	0.2
Deer mouse	25	0.6	42	3.0
Masked shrew	40	1.6	25	3.0

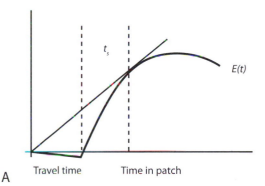

A predator foraging in a patch will eventually deplete the food available in that patch, and the encounter rate for prey will drop over time. Thus, if we plot the net caloric intake over time $E(t)$, we would expect a graph similar to Figure 17.3.

In Figure 17.3A, the forager gains a lot of energy after it enters the patch (i.e., the curve is rising sharply), but as the forager spends more and more energy searching for fewer and fewer prey, the curve begins to decline. The curve in part B is more realistic because it includes the travel time to get to the patch. Notice that travel has a cost associated with it; the first part of the graph dips below the horizontal line, indicating net energy lost while traveling.

When should the predator quit foraging in this patch and move on to another patch? You might assume that the predator should leave when the energy gain is maximized (i.e., at the highest point on the curve), but this is incorrect. The reason is that the forager is presumably trying to maximize its average net caloric intake rate E/T, but in this case T includes both travel time between patches and time spent within each patch. In other words, we are looking for the average rate of change. The maximum average net caloric intake rate is then the slope of the tangent line from the origin (Figure 17.4).

If the forager quits the patch at time t_1 in Figure 17.4, then the slope of the line from the origin to that point on the curve is less than the slope of the tangent line at time t_2. Since the slope of the line to any point on the curve is the average net caloric intake rate at that point in time, it should be obvious that E/T is maximized at the point on the curve where the tangent line passes through the origin (for details, see Kolmes et al., 1997). This is called the marginal value theorem; it was borrowed from economics and first applied to behaviors by Charnov (1976).

The patch model can be tested by manipulating the travel time between patches. For example, if travel time between patches is lengthened, then it becomes costly to move between patches and the forager should remain longer in the patch before quitting and moving on to a new patch (Figure 17.5). This is illustrated by the lower slope of the tangent line in part B of Figure 17.5. You will test some of the predictions of optimal foraging theory in the exercises that follow.

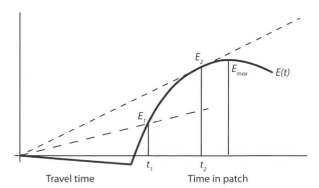

Figure 17.4 The average net caloric intake rate at three points in time. The E/T is maximized at the tangent to the curve.

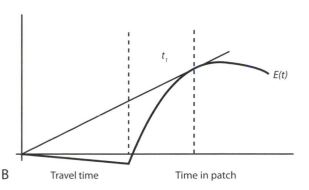

Figure 17.5 Two graphs of travel times between patches of equal value. Travel time is short in (**A**) and longer in (**B**).

Exercise 1: Profitability and Prey Choice

In this exercise, you will present food items of varying energy content and handling times to foraging mammals. You will use squirrels as foragers (predators) and various nuts as prey.

1. Your instructor will provide you with two types of peanuts for your first trial. In this trial you will give the squirrels a choice between peanuts in the shell and peanuts with the shells removed. Because peanuts usually come two nuts to a shell, you will have to carefully cut the peanuts in the shell in half so that they contain a single nut. Why is this a critical step? Peanuts in the shell have greater handling time because the squirrel must remove the shell to eat the energy-rich nut inside. Therefore, you would predict that the squirrels should prefer the more profitable shell-less peanuts over the shelled peanuts.

2. Locate a place where squirrels are common and easy to observe.

3. Place a two-foot-square whiteboard near the squirrel's center of activity. (Painting the tray white makes it much easier to observe when nuts have been removed.)

4. On the whiteboard, place an equal number of shelled and shell-less peanuts.

5. Locate a place where you can sit and easily observe your subject squirrels but will not disturb their behavior.

6. Using binoculars, if needed, record how long each squirrel remains at the tray (patch), what nuts it eats or removes from the patch, and the order in which the two types of nuts are chosen.

7. Repeat the process several times to ensure that you have enough replicates.

8. Summarize your results graphically, compare the removal rates of shelled and shell-less peanuts statistically, and write a short methods and results section in journal format (your instructor will provide details and resources).

9. Develop a new study that tests whether squirrels forage optimally. Think about other ways to vary handling time or energy content of the food choices.

10. Create a hypothesis grounded in optimal foraging theory and a prediction about how squirrels should forage. Develop a study design that can test this prediction. Consult with your instructor before carrying out the study.

11. Conduct the study, and carefully record your data.

12. Present your study design and results to the class.

Exercise 2: Foraging in Patches

In this experiment, you will give squirrels a choice of two patches, a poor-quality patch and a high-quality patch. Patch quality will be set by the density of nuts. Each nut is assumed to be of equal caloric value.

1. Your instructor will provide each team with two feeding trays (each tray is two feet square with a two-inch-high perimeter; trays are filled with sand).

2. Count out 50 peanuts, and randomly bury them in the sand in the first tray; this is the high-quality patch.

3. Count out 10 peanuts and randomly bury them in the second tray; this represents the poor-quality patch.

4. Create a set of predictions for foraging time in patch, patch quitting time, and rate of foraging within each patch.

5. Set out the two patches within five to six feet of one another near the center of squirrel activity.

6. Use binoculars and a stopwatch to record the squirrels' foraging behavior and times.

Exercise 3: Foraging with Risk

This experiment will test the foraging efficiency of squirrels under varying predation and competition risks. This experiment is similar to that of Newman and Caraco (1987), and it is a good idea to read that paper before you begin. The basic premise is that patches that are out in the open or far from cover are riskier places to forage because of the risk of predation. However, even patches near cover can be risky if there are many aggressive competitors (e.g., other squirrels) all trying to forage on the same food patches.

There are essentially four variables:

- **Squirrel feeding rate**: In this experiment, you use sunflower seeds because squirrels are more likely to eat them on the spot instead of caching them for later. You will record the number of seeds eaten per minute by each squirrel. Record the number of seeds and seconds even if the squirrel leaves the patch before a minute is up. If more than one squirrel is foraging in the patch at one time, record

Table 17.2 Summary data table for squirrel foraging study

Distance to cover =			Patch size =		
Group size	Mean feeding rate	Standard deviation of feeding rate	Mean group size	Standard deviation of mean group size
1				
2				
3				
4				

each squirrel's feeding rate separately and the group size (see below). If group size changes during the minute, record the average group size for that minute. If a single squirrel stays for more than one minute, record it again for another minute (up to five minutes).

- **Group size**: Group size is the number of squirrels foraging at the same time in the same patch (or nearby).
- **Distance to cover**: This is the distance in meters to tree cover. The predation risk, energy costs, and travel time all increase with distance to cover. You will place your food patches at either 5 or 15 meters from suitable cover.
- **Patch size**: The number of seed-containing trays in each patch will be either one tray (small patch) or four trays (large patch).

Before you begin, think about the following questions and generate predictions that can be tested from the data you collect for the four variables just discussed:

- Would a squirrel prefer to forage in a large group or alone?
- Would group size likely be larger at greater distances to cover? Why?
- Given identical food items, would feeding rates increase or decrease with distance to cover?
- How might patch size and distance to cover interact to affect group size and/or feeding rate?

Your instructor will divide the class into groups and assign each group a patch type (small or large) and a distance (5 or 15 meters). One person in each group will be responsible for recording the data onto the data sheets, while the others will report the feeding rates and group sizes every minute (a sample data table appears in the Appendix to this chapter). They should also note any aggressive behaviors (e.g., chasing).

1. Place your patch at the appropriate distance from tree cover, and put an equal number of sunflower seeds in each tray.

2. Retreat to a place where you can observe but not disturb the squirrels, and begin observations (be patient; you may have to wait a bit until the squirrels find the patch).

3. When you are finished, remove your patches and all seeds from the site. Summarize the raw data from your data sheet by calculating the means and standard deviations as shown in Table 17.2. Class data will have to be pooled from these summary sheets prior to statistical analysis.

4. For feeding rate data, calculate a mean and standard deviation for each group size for which there are observations. For group size observations over time, calculate an overall average group size across all times.

Now you will look for a relationship between the squirrel's feeding rate and (1) group size, (2) patch size, and (3) distance to cover. To do this, create a scatterplot of the raw data (not the means), plot a "best-fit" line to that data, and provide the equation for that line with a constant slope. A linear regression determines the relationship between a dependent variable and an independent variable. Linear regression assumes (1) that there is a plausible reason to assume a cause-and-effect relationship between the variables, (2) that the data were randomly collected, and (3) that observations are independent (if you collected five sets of feeding rates all from the same squirrel, then these sets are not independent). The regression equation is the familiar equation for a line $y = mx + b$, where x and y are the data points, m is the slope of the line, and b is the y-intercept. The regression equation has a constant slope m, meaning that the feeding rate changes at a constant rate as group size (or other variable) changes. The R^2 value is the percentage of all the variance in one variable that can be accounted for by the linear regression model. If R^2 was 0.87 for a plot of squirrel-feeding rates (y) versus group size (x), then 87% of the variance in feeding rate can be accounted for by group size. It also means that 13% of the variance remains unaccounted for. A p-value is also reported in most analyses of linear regressions. The p-value relates to the null hypothesis. In a linear regression, the null hypothesis is that there is no relationship between the variables. If true, the regression line would be horizontal (slope = 0.0). If the *p*-value is less than 0.05, we can conclude that the null hypothesis is unlikely to be true and that the observed relationship is not due to chance. Here are the steps to follow:

1. Use Excel, OpenOffice, or a statistics program (specified by your instructor) to plot the regressions of the variables you collected in your squirrel study.

2. Consider each regression carefully. What patterns do they suggest?

Regressions are useful for determining the relationship between two variables (assuming the relationship is linear and not on a curve). However, if you want to know whether the mean feeding rate for squirrels at large patches is different from those feeding at small patches, you need a different statistical test—an analysis of variance (ANOVA). You could use a student *t*-test, but an advantage of ANOVA over *t*-tests is that it can (1) test more than two means simultaneously, and (2) determine the effects of more than one factor and any interactions between them. Suppose you want to test the effects of distance to cover and patch size on feeding rate, and you also want to know whether there is any interaction between patch size and distance to cover. In this case, you use an ANOVA.

Your instructor will provide you with additional instructions on how to perform ANOVAs on the statistical platform at your institution. Use your ANOVA analysis to answer the following questions about your study:

- Do your results support or falsify your initial hypothesis?
- Which patch size gives the highest feeding rates?
- Do feeding rates increase or decrease with distance from cover?
- How do distance from cover and patch size interact?
- How does group size change with distance from cover?
- How does group size change with patch size?
- Why might your data deviate from your predictions?
- What selective factors may be at play in shaping the squirrels' foraging behaviors?

Appendix

Table 17.3 Sample data table for squirrel foraging project

Distance to cover: 5 or 15 meters		Patch size: Small or large	
Time	Feeding rate (# of seeds eaten/minute)	Group size at time X	Aggressive behaviors or comments

18   Field Karyotyping

Time Required
- One 3- to 4-hour lab period for each of the exercises

Learning Objectives
- Understand the basics of mitosis and chromosome structure.
- Understand how karyotypes are made.
- Understand what kinds of information karyotyping can yield.
- Learn how prepare a field karyotype.
- Learn how to collect and preserve tissues for DNA analysis.

Equipment Required
- Mitotic inhibitor (sterile vinblastine sulfate, USP), Eli Lilly and Company
- Eagles growth medium (GIBCO Laboratories)
- Disposable 15-ml centrifuge tubes with caps (Corning Laboratory)
- 0.075 M potassium chloride solution
- Carnoy's fixative
- Giemsa blood stain
- Dehydration solutions (acetone and xylene)
- 6-inch Pasteur pipettes
- Rubber latex bulbs
- 1-cc insulin syringes
- Coplan staining jars
- Microscope slides (1.0 mm) and storage boxes
- Coverslips (24 by 50, No. 1 thickness)
- Permount slide mounting medium
- Clinical centrifuge, or hand-cranked field centrifuge
- Gasoline power generator (for clinical centrifuge)
- Centrifuge tube racks
- 100-ml graduated cylinder
- Beakers

Background

Mammalogists are often interested in questions that require an understanding or comparison of the chromosomes of one or more populations or species. Understanding the chromosomal complement of a mammal is useful for studying taxonomic relationships, chromosomal aberrations, and cellular function and for understanding past evolutionary events. For example, the duck-billed platypus (*Ornythorynchus anatinus*, in the Monotremata order) was discovered to have five pairs of sex chromosomes instead of the single pair (XX or XY) found in most mammals (Grutzner et al., 2004). Surprisingly, platypus sex chromosomes appear homologous with bird sex chromosomes and not with other mammals. This implies that therian sex chromosomes evolved (probably from an autosomal pair) after monotremes 166 million years ago (Veyrunes et al., 2008).

The chromosomal complement, or karyotype, is typically uniform within a species. However, there may be considerable karyotypic variation among populations of a species. Thus, karyotypes often aid in our understanding of the relationships among species. For example, close examination of chromosomes reveals that rearrangement of chromosome segments into different combinations can explain much of the variation in species karyotypes (for a review see Ferguson-Smith and Trifonov, 2007).

What Is a Karyotype?

A karyotype is either (1) the number and appearance of chromosomes in a eukaryotic cell, or (2) the complete set of chromosomes in an individual organism (or a species). In essence, karyotypes describe the number of chromosomes and what they look like (i.e., their length, centromere position, banding pattern, and any other physical differences). The chromosomes are arranged in pairs by size (largest first) and, for similarly sized chromosomes, by position of centromeres, with the sex chromosomes positioned last (Figure 18.1). This standard format is known as a karyogram (or idiogram). The short arm, called "p" (for petite), is always at the top, and the long arm, designated "q" (because it follows letter "p"), is always at the bottom.

Karyograms illustrate the number of chromosomes in an individual's or species' somatic cells; this somatic number is referred to as 2n (diploid). Gamete cells have

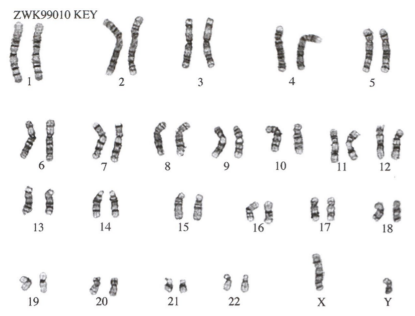

Figure 18.1 A karyotype of a human male showing the 23 homologous pairs of chromosomes (including an XY pair of sex chromosomes) stained with Giemsa. *Reprinted with permission of University of Wisconsin Cytogenetics–Wisconsin State Laboratory of Hygiene. Copyright UW Board of Regents.*

a chromosome number of n (haploid). For example, in humans 2n = 46, where n = 23.

Scientists typically read chromosomes in five ways. When scientists compare chromosomes from different species (or individuals from different populations), they look for differences in the following characteristics:

- **Size**: Chromosomes can vary in length by up to 20 times between genera within a family. Such size differences probably result from differences in the number of DNA duplications. Alternatively, differences in relative size of chromosomes are caused by interchange of unequal lengths of segments of chromosomes.
- **Centromere position**: Centromeres are required for normal chromosome separation during mitosis or meiosis. Normally, each chromosome has a single centromere located near the middle of the chromosome. Such a position is termed metacentric. Other chromosomes may have their centromere positioned near the very end of the chromosome (acrocentric) or off-center (submetacentric). Differences in the centromere position occur by translocations (rearrangement of parts between nonhomologous chromosomes).
- **Chromosome number**: Differences in the basic number of chromosomes may occur owing to repeated translocations that remove essential DNA

from a chromosome by adding that DNA to other chromosomes. This eventually allows the loss of a chromosome without harm to the organism. For example, great apes have one more pair of chromosomes than humans, but the essential great ape genes were added (translocated) to other chromosomes in humans.

- **Satellite position and number**: Some chromosomes have an additional constriction (besides the centromere) near the distal tip of the chromosome called a satellite. Satellites are small fragments attached to the main chromosome by a strand of chromatin. Differences in number and position of satellites can be useful in cytogenetics.
- **Heterochromatic banding patterns**: Heterochromatin refers to regions of densely packed DNA that stains darker than euchromatin (lightly packed regions). In the karyotype in Figure 18.1, Giemsa staining reveals light and dark bands (G-banding). Giemsa stains phosphate groups of DNA. Each chromosome has a characteristic banding pattern, but both members of a chromosome pair have identical banding patterns. Thus, differences in the relative staining of these regions can be useful when comparing karyotypes. A common misconception is that bands represent single genes, but in fact the thinnest bands contain more than a million base pairs and potentially hundreds of genes.

A full comparison of karyotypes may reveal variation between sexes, members of a population (polymorphisms), geographic regions (geographic races), or individuals (abnormal individuals).

How Are Karyotypes Produced?

Conventionally, karyotypes were made by injecting the subject animal with mitotic inhibitors such as colchicine citrate or vinblastine sulfate, which arrests cell division at metaphase when chromosomes are condensed. After a 24-hour period the animal is sacrificed, and the bone marrow is removed and placed on a microscope slide. This technique, which uses live animals, has been modified for use with postmortem samples (dead animals). The postmortem technique described below is more appropriate for field work where housing live animals overnight is difficult or impossible.

Be sure to follow the procedures spelled out in your approved Institutional Animal Care and Use Committee (IACUC) protocol and the Guidelines of the American Society of Mammalogists for the use of wild mammals in research, which can be found at http://www.mammalsociety.org/.

Exercise 1: Field Karyotyping

Following the capture of an animal (see Chapter 6), the animal is euthanized (alternatively, animals euthanized as part of another study can also be used) and the following information is recorded:

- Sex: male/female
- Age: adult/immature
- For females, record whether lactating or pregnant
- Weight to nearest 0.1 g
- Head + body length to nearest 1 mm
- Tail length to nearest 1 mm
- Length of right hind foot to nearest 0.5 mm

1. Immediately after euthanasia, expose and remove the femur. Cut the head of the femur off near its proximal tip to expose the bone marrow cavity.

2. Using a narrow-gauge needle attached to a syringe filled with Eagles medium, insert the needle into the marrow cavity and flush the marrow contents into a 5-ml disposable tube.

3. Gently break up any marrow clumps with a 23-gauge needle (or by drawing the marrow solution back and forth into the syringe).

4. To each 5 ml of media-marrow suspension, add 0.1 ml of a 0.001% colchicine solution (mitotic inhibitor) or 0.01% Velban (sterile vinblastine sulfate), shake, cap the tube, and incubate at 37°C for 90 minutes. Under field conditions, incubate samples close to one's own body.

5. Centrifuge the tube for 3 minutes at approximately 3,000 revolutions per minute (rpm) using a manual (hand-cranked) or battery-operated centrifuge until a pellet forms in the bottom of the tube.

6. Carefully remove and discard the supernatant (leaving the cell pellet), and add approximately 3 ml of hypotonic solution (0.075 M potassium chloride).

7. Re-suspend the cells in hypotonic solution by aspirating the cell pellet with a pipette.

8. Recap the tube, and incubate at 37°C for 15 minutes.

9. Centrifuge the tube for 3 minutes at approximately 3,000 rpm.

10. Carefully discard the supernatant (leaving the cell pellet), add approximately 3 ml of fixative solution, and re-suspend the cells in fixative. Carnoy's fixative is one part glacial acetic acid and three parts absolute methanol (made fresh before use). This is Wash 1.

11. Centrifuge for 3 minutes at approximately 3,000 rpm.

12. Discard the supernatant fixative, add approximately 2 ml of new fixative, and re-suspend the cells.

13. Aspirate the entire cell suspension in a pipette and filter through cheesecloth into the original centrifuge tube. If necessary, add more fixative to bring the volume of suspension to approximately 3 ml. This is Wash 2.

14. Centrifuge for 3 minutes at approximately 3,000 rpm.

15. Discard the supernatant fixative, add approximately 1 ml of fresh fixative, and re-suspend the cells. The solution should be visibly cloudy.

16. Place a clean microscope slide on the benchtop. Allow 3-5 drops of the cell suspension to drop from a distance of 1-2 feet above a clean microscope slide. This allows the chromosomes to spread out a bit and makes them easier to see. Allow the preparation to air dry.

17. Insert the microscope slide into a Coplin jar containing Giemsa stain for 15 minutes. Giemsa stain is one-part stock Giemsa with eight parts hot tap (or distilled) water (made fresh before use).

18. Remove the slide from the Giemsa, and pass it through a series of dehydration baths. Dehydrate quickly, using 2 dips, 1 second each, in two baths of acetone, one bath of acetone and xylol (1:1), and two baths of xylol.

19. Coverslip immediately with 3-4 drops of Permount.

Exercise 2: G-banding Chromosomes with Trypsin

This protocol may provide superior G-banding results because it uses an enzymatic treatment of the slides. Investigators should adjust the times to suit the slide quality desired. This exercise begins after the slide has air dried in step 16 above.

1. Age the air-dried slide overnight in a drying oven at 55° to 60°C. Remove the slide and bring to room temperature just before banding.

2. Grasp the slide with forceps, and immerse it in a Coplin jar containing the 0.025% trypsin working solution for 8 to 10 seconds, moving the slide back and forth

gently. Trypsin working solution 0.025% is 0.5 g trypsin (Difco, 0.025% w/v final) in 200.0 ml Earle's balanced salt solution. Store up to 1 day at room temperature.

3. Briefly rinse the slide in 1% FBS to inactivate the trypsin.

4. Pre-rinse the slide by dipping in Gurr's buffer, using the same agitation technique as in step 2. To make Gurr's buffer solution, pH 6.8, dissolve 1 Gurr's buffer tablet in 1 liter sterile H_2O.

5. Place slide in a Coplin jar containing Giemsa staining solution for 8 to 10 minutes. Make Giemsa staining solution by adding 1 ml Giemsa stain (Azure Blend,

Harleco 620G/75, from EM Science; 2% v/v final) to 50 ml Gurr's buffer solution, pH 6.8; store up to 2 months at room temperature.

6. Rinse slide in sterile distilled water until the stain no longer discolors the water, using the same agitation technique as in step 2.

7. Allow slide to air dry. Examine by light microscopy using a phase-contrast microscope to determine the quality of banding. Adjust trypsin exposure or duration of staining as required.

8. Once the optimal banding quality has been achieved, coverslip and analyze the slides.

Exercise 3: Analyzing the Karyotype Manually

Digitally photograph appropriate spreads, produce 8-by-10-inch high-contrast photographs of your chromosome spreads, and print these images. Cut each chromosome from the photograph and arrange the chromosomes according to size and position of the centromere (this can also be done digitally using Adobe Photoshop). To avoid losing one, it is best to do this one at a time rather than cutting them all out at once and then trying to match them.

This process may be difficult if the chromosome spreads on the slides are of poor quality. In this case, use the sample human chromosome spread in the Appendix A (provided courtesy of the Wisconsin State Laboratory of Hygiene). Carefully cut out each chromosome, and compare each chromosome to its potential homolog in partially completed human karyotype in Appendix B.

Tape or glue each chromosome next to its pair on a form supplied for this purpose. The analysis involves comparing chromosomes for their length, the placement of centromeres (areas where the two chromatids are joined), and the location and sizes of G-bands.

Determine the diploid number of chromosomes, sex, and number of metacentric, submetacentric, and acrocentric chromosomes. If possible, compare the karyotypes from several mammalian species and note any differences in the karyotypes. Analyzing karyotypes by hand is time consuming. It is now easier to do it using image analysis software on a computer.

Exercise 4: Measuring Chromosomes with ImageJ Software

You need to create digital images of the karyotypes from your preparations in Exercises 1 and 2. Your instructor will show you how to use the digital camera attached to your microscope. Alternatively, digital images of human karyotypes can be found on the web.

Before you begin taking photos of your chromosome preparations, you will need to take a photograph of a micrometer slide to set the scale of your images. A stage

micrometer has a finely divided scale etched on the surface (Figure 18.2). The scale is a precise true length and is used for calibration.

Place the micrometer slide on the microscope stage. Under the lowest magnification, adjust the focus so that the etched lines are in focus. Take a digital photo and save it with an appropriate title (include the magnification in the title—e.g., micrometer 4X). Repeat this procedure for the other objectives you plan to use (when switching objectives be careful not to scratch the lens across the micrometer slide).

Remove the micrometer slide, and replace it with the chromosome preparation slides you wish to measure. Locate regions of interest, and photograph and save them. Record each image with a unique and informative label, and save them to a folder (include the objective used). Be aware that each microscope is slightly different, and it is good practice to take micrometer photos at each power for each microscope.

You will apply image analysis techniques to your digital images, using the public domain software ImageJ, written and maintained by Wayne Rasband at the National Institute of Mental Health, Bethesda, Maryland, USA (ImageJ is the successor of NIH Image). ImageJ can be run on any system for which a Java runtime environment (JRE) exists (e.g., Windows, Mac, Linux).

If you have not already done so, go to the ImageJ website and download the appropriate version for your

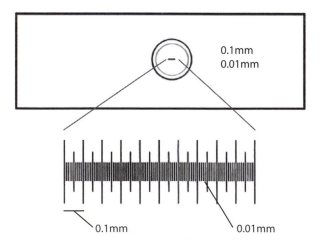

Figure 18.2 A drawing of a stage micrometer (*top*) showing the scale ruler as the center and the dimensions of the large and small lines etched on the slide (*right top*) and a close-up view of the scale (*bottom*).

operating system: http://imagej.nih.gov/ij/. Java applications will only use the memory allocated to them. Under Edit Options > Memory & Threads, you can configure the memory available to ImageJ. The maximum memory should be set to three-quarters of the available memory on your machine.

In the first part of this exercise you will use the photo of the micrometer slide to create a scale bar for each subsequent image. Suppose you are using the photo of a micrometer slide taken at 40X.

1. Begin by opening ImageJ, and then open the image file for the 40X micrometer slide taken earlier (i.e., File > Open, and then browse to the correct image).

2. Select the straight-line tool from the ImageJ toolbar (fifth from the left, Figure 18.3), and place the cursor (which is a crosshair) carefully at the start of one major line, click and drag the crosshair to the next major line (or over several major lines). Recall that the distance between major lines is recorded on the surface of the micrometer slide (the larger of the two numbers). For example, if your micrometer slide is marked 0.1mm and 0.01mm, then the major lines are each 100 microns apart and the smaller lines are each 10 microns apart. Record the length in pixels for the distance you measured (it is given just under the ImageJ toolbar in pixels). For example, if there are 1865 pixels in 0.1 mm (100 microns), you would record 1865/100 μm (or 18.6 pixels per micron) in your notes. Repeat this for each magnification.

3. In ImageJ go to Analyze > Set Scale. Enter the distance in pixels from your measurement (i.e., 1865 in this example) and the known distance (i.e., 0.1 in this example) and the unit of length (i.e., mm). Check the "Global" box to apply this scale to all images in the current ImageJ session. If you restart the program, you have to re-enter the scale each time, so be sure to have all your ratios written down.

4. Click OK, and test the settings by re-measuring the major lines and making sure they are 100 microns apart. (Note: Use the keyboard shortcut Ctrl+m.)

5. Open your first chromosome image (File > Open). Add a scale bar to the image by going to Analyze > Tools > Scale Bar. Specify the width, thickness, font size, color, and location of the scale bar, and click OK. Now you are ready to measure some chromosomes. You can do this in two ways. The first way is to use the tools already present in ImageJ to measure each chromosome. Set the scale as described above using Analyze/Set Scale.

- Use the magnifying tool to enlarge the region of interest. Click on the line segment tool to select it. (To use the line segment tool, right-click on the straight-line tool—fifth from left in tool bar.)
- Moving the mouse cursor over the chromosome, click and drag from one tip to the middle; click there once. Then continue moving the crosshair

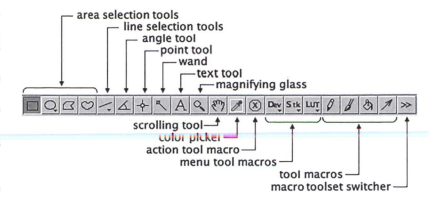

Figure 18.3 A view of the ImageJ toolbar with the tools listed.

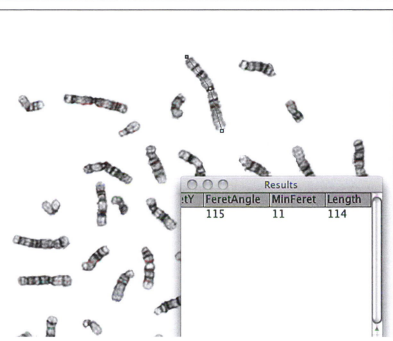

Figure 18.4 The ImageJ window showing the icon for the line segment tool in the toolbar, a chromosome with a line segment over it, and a results window showing the length of the chromosome as 114 microns.

to the other tip and double-click. You should now have a line with two segments that runs the length of the chromosome (Figure 18.4).

- Use the Analyze/Measure command to measure the length of the chromosome line. The measurement appears in the Results window under length (see Figure 18.4). Measurements can be transferred to a spreadsheet by right-clicking in the Results window, selecting Copy All, switching to the spreadsheet program, and then pasting.

6. The second way to measure chromosome arm lengths is to use a plugin written specifically for this task called LEVAN, developed by Yugo Sakamoto and Adilson

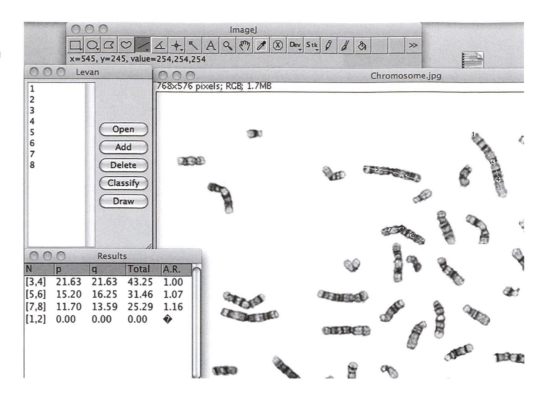

Figure 18.5 The Levan plugin window in ImageJ showing several marked chromosome arms and the results table.

Zacaro and available at http://rsb.info.nih.gov/ij/plugins /levan/levan.html. Visit this site, download the Levan.jar file, and move it to the plugins folder in ImageJ, then re-start ImageJ, and the Levan command will appear in the drop-down list under Plugins in the menu bar.

- Close any open images. Go to Plugins > Levan to open the Levan plugin in a new window.
- Click the Open button and browse to the image of your chromosomes. Use an image where there is little or no overlap of chromosomes. Alternatively, use one of the processed images available from the web.
- Levan opens the karyotype image in a new window. Choose the line selection tool on the toolbar, and with the mouse trace each arm of each chromosome. As you do this, each arm will be added to the list in the Levan window.
- Click on Classify in the Levan window to classify each chromosome as metacentric, submetacentric, etc. Levan also gives you the length of the p and q arms in pixels. You will need to convert this to microns manually.
- Click on the Draw button in the Levan window to add the numbered arms to the photo (Figure 18.5).
- To save the data to another location, go to File > Save As in the ImageJ menu.

Appendix A

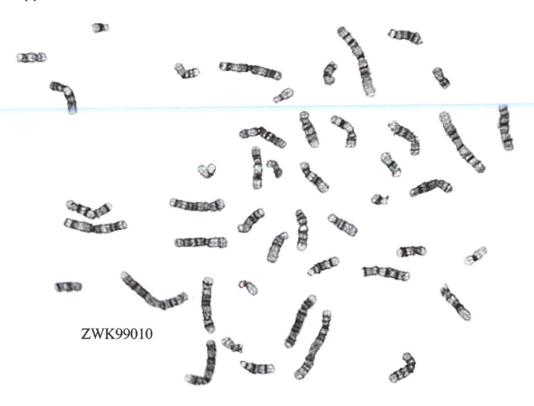

ZWK99010

Figure 18.6
A human chromo-some preparation stained with Giemsa. *Reprinted with permission of University of Wisconsin Cytogenetics–Wisconsin State Laboratory of Hygiene. Copyright UW Board of Regents.*

Appendix B

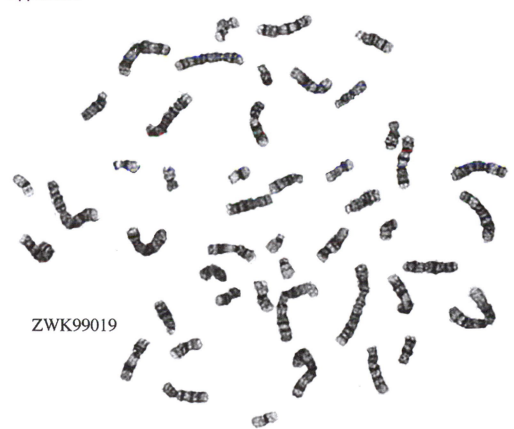

ZWK99019

Figure 18.7
A second human chromosome preparation stained with Giemsa. *Reprinted with permission of University of Wisconsin Cytogenetics–Wisconsin State Laboratory of Hygiene. Copyright UW Board of Regents.*

19 Non-invasive Hair Sampling

Time Required
- One 3- to 4-hour lab period for each of the exercises

Learning Objectives
- Understand the basics of hair morphology.
- Understand how hair samples can be collected in the field.
- Understand how reference collections are made.
- Learn how to analyze hairs using ImageJ software.
- Learn how to preserve hair samples for DNA analysis.

Equipment Required
- PVC tubing of various diameters
- Double-sided sticky tape or rodent glue boards
- Raw meat (bait) for every trap station
- Thumbtacks (handful)
- Hair sample envelopes (~20)
- Permanent markers
- Cordless drill/screwdriver and exterior wood screws
- GPS with location of the station
- Wooden boards (two per hair station)
- ImageJ software (free open source)
- Microscope slides and coverslips
- DPX or Permount mounting medium
- Reagents and stains (described below)
- Nail polish (clear)
- Microscope
- Coplin jars
- QIAamp DNA kit
- Microcentrifuge
- Adjustable micropipettes and disposable tips
- Vortex shaker

Background

Mammalogists are often interested in surveying the distribution and abundance of species within a region. Although it may be possible to livetrap and identify individuals, it is often time consuming and expensive if the area is large or if the species are rare or elusive (e.g., many carnivores). Non-invasive techniques have been developed for carnivores (Long et al., 2008) and small mammals (Pocock and Jennings, 2006) as alternatives to traditional trapping. These non-invasive techniques include remote camera systems (see Chapter 12), the use of track plates, and hair traps. Each method has its pros and cons, and combinations of methods are needed to meet some study objectives.

Hair sampling has gained popularity as a non-invasive method for sampling species distribution or occupancy (see Kendall and McKelvey, 2008). In addition, hair sampling combined with molecular techniques is used to measure population density, monitor mammal populations, and increasingly to develop management strategies for wide-ranging or elusive species.

Hair traps are now routinely used in large-scale field surveys whose goal is to detect small to medium-sized terrestrial mammals. Hair traps come in a variety of shapes, sizes, and configurations. They are described as hair snares, hair tubes, sticky traps, and other names depending on the specific design. For example, hair tubes require an animal to enter a tube and pass by sticky tape, which collects a hair sample, before reaching bait at the far end of the tube. Hair snares, by contrast, work by collecting hairs on a wire or stiff brush as the animal rubs against it.

Regardless of the method used to collect the hair samples, the hairs left behind can be analyzed by microscopic examination or DNA methods to reveal the species that deposited the hairs.

Hair Morphology

Hair is a diagnostic character for the Mammalia. It is associated with the integument and grows out of a structure called a follicle (see Vaughan et al., 2015). The root of the hair lies within the follicle, and the portion above the skin surface is the hair shaft. The bulb forms the base of the root. Structurally, hair is composed of interconnected keratin proteins in the form of long fibrils. Keratin makes hair highly resistant to wear and chemicals. This is because keratin proteins have many cross-linkages and disulfide bonds (S-S) link adjacent protein chains. In essence, a hair is a long thin tube (Figure 19.1).

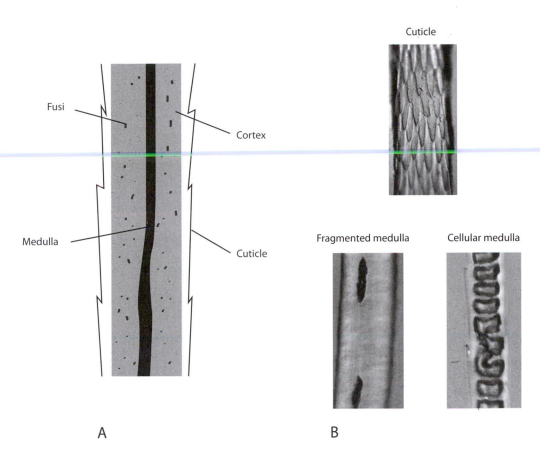

Cuticle

Fusi

Cortex

Medulla

Cuticle

A

Fragmented medulla

Cellular medulla

B

Figure 19.1 (**A**) A diagram showing the internal anatomy of a hair cut lengthwise. (**B**) Three photos of mammalian hairs. The cuticle on the top is from a mink, the fragmented medulla on the left is from a brown bear, and the cellular medulla on the right is from a chinchilla. *Photos courtesy of Russ Crutcher at www.microlabnw.com.*

The hair shaft consists of a translucent outer layer called the cuticle, consisting of overlapping cells that resemble the scales on a snake. These cuticle cells are keratinized (hardened) and contain no nuclei and no pigments. The scales overlap and have their free end pointing distally. Cuticle patterns can be helpful in identifying species, but these patterns may vary even within an individual depending on the location of the hair (e.g., a guard hair may have a different cuticle pattern from underfur).

Just inside the cuticle is the cortex. It is composed of long, keratinized filaments running parallel to the hair shaft. Trapped within these filaments are air spaces called fusi. Hair color is primarily due to the density of pigment granules (melanin) in the cortex. At the very core of the hair shaft is the medulla. The medulla varies from individual to individual and between hairs of a given individual. The medulla may be continuous throughout the entire hair shaft, fragmented, or even absent altogether (Figure 19.2).

Many mammalian hair shafts show variation in diameter along their length. For example, the base may be smaller in diameter than the mid-section, and the tip may taper to a blunt point. A number of descriptive terms have been developed to aid in the analysis of hairs by forensic scientists and wildlife biologists. For example, Figure 19.2 illustrates that the medulla may be fragmented, interrupted, continuous, or absent. Likewise, there are terms to describe the cuticular scale patterns (Figure 19.3).

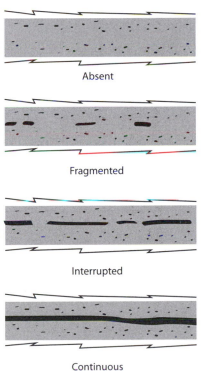

Absent

Fragmented

Interrupted

Continuous

Figure 19.2 Diagrams of longitudinally cut hairs showing differences in the medulla.

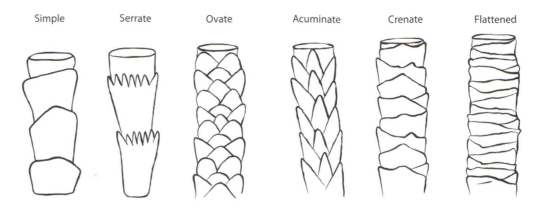

Figure 19.3 Drawing of the cuticle morphology of mammalian hairs.

| Simple | Serrate | Ovate | Acuminate | Crenate | Flattened |

Exercise 1: Field Methods for Collecting Hairs

A large number of devices are available for collecting hair in the field. Among the most popular are hair tubes of various kinds and hair snares. Hair snares use barbed wire or metal brushes (e.g., gun-cleaning brushes) to snag the hairs as an animal rubs against the snare. Hair snares are usually used for larger mammals such as ungulates and carnivores. For small mammals, the most commonly used device is a hair tube fitted with sticky tape. Hair tubes are baited and set out in areas where the target species are likely to occur. Hair tubes are made in a variety of shapes and sizes.

In this lab, you will make your own hair tubes and set up a collecting transect for surveying small mammals such as shrews and rodents. Hair tubes for small mammals need to be made in various sizes. Cut lengths of PVC plastic pipe into 12-inch (30-centimeter) lengths. Use pipes of varying diameters to cover the range of small mammal body sizes you expect to encounter. For example, cut one length each of 4-inch-, 3-inch-, 2-inch-, and 1.5-inch-diameter PVC. Line the inside wall on one side with one or more strips of double-sided carpet tape, leaving one side of the pipe uncovered to act as the floor. It is useful to cut the PVC tube in half lengthwise and then use duct tape to put the two halves back together (Figure 19.4). This allows easy access to the interior of the tubes when removing hair samples and replacing sticky tape in the field. Cut wooden plugs to fit into the PVC at one end. These plugs are fitted on one side with wire mesh tea-infuser balls that hold the bait. Traps are baited with a combination of rolled oats, honey, and peanut butter. The ends of the PVC are plugged at the site with the bait facing the inside of the tube. Baited hair tubes are placed next to each other on the ground near potential runways (Figure 19.4).

Small Carnivores

The following hair-sampling device is modified from those developed by Mowat and Paetkau (2002) for the American marten (*Martes americana*). The traps consist of two pine boards joined together lengthwise to form a roof. Each board is 1 × 6 × 24 inches (2 × 14 × 60 centimeters); the two are glued and screwed together along their long axis to form a 90° angled roof (Figure 19.5). Short strips of double-sided carpet tape or Catchmaster rodent glue boards are attached to the underside of the roof. Glue strips are attached near both ends of the trap; the center section is used for bait or scent lures.

The trap is either screwed to a tree trunk at chest height or placed on the ground in suitable cover. The traps form a triangular tunnel through which the animal

Figure 19.4 Hair tubes made of PVC pipe. (*Left*) Hair trap strapped to a tree branch with Velcro straps. (*Right*) A set of three hair tubes of different diameters set along a log.

Figure 19.5
Wooden hair trap used for small carnivores. Sticky tape is affixed to the underside of the wood. (*Left*) A hair trap set vertically on a tree trunk. (*Right*) A hair trap set horizontally along a log

can enter. Traps are baited with raw meat or other suitable bait for the target species.

Deploying and Checking Hair Tubes

Place 20 sets of PVC hair tubes (a set includes all diameters) in suitable habitat along a transect. Sets should be placed at roughly 20-meter intervals, numbered, and flagged. The number of tubes used for each site in any survey should be standardized. Attach traps to vertical tree trunks using screws or on horizontal branches using Velcro tape or hose clamps around the trap and branch (Figure 19.4).

Carefully check the wooden carnivore traps by unscrewing them from the tree and examining the glue boards for hair. If a board has hair, remove the entire glue strip (or tape) and place it in a sample bag or vial labeled with the station, location, your initials, date, time, and trap position (tree trunk, tree branch, ground, etc.). Check the hair tubes by removing the wooden plugs and duct tape to separate the two halves. Remove the tape and replace it (if needed). Check all trap stations for bait, and replace the bait if necessary. Reassemble the traps, and put them back in their previous locations.

Exercise 2: Creating a Hair Reference Collection

Creating microscope slides of hair samples is relatively easy. The basic setup includes several Coplin jars (i.e., jars that hold slides while they are staining or rinsing), microscope slides, coverslips, and mounting material to attach the coverslips (e.g., Permount or DPX mounting medium) (Figure 19.6).

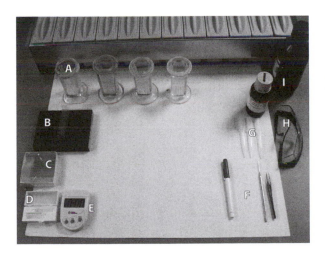

Figure 19.6 Typical staining setup: (**A**) Coplan jars, (**B**) slide rack, (**C**) microscope slides, (**D**) 40 mm cover slips, (**E**) digital time, (**F**) pen and forceps, (**G**) disposable pipettes, (**H**) safety glasses, (**I**) coverslip mount (e.g., DPX Mountant or Permount).

Two types of preparations are necessary. The first is a cast of the outer cuticle that will be used to identify the cuticle scale type. Casts are made by placing the guard hair in a semi-dry bed of clear nail polish and then removing it when the nail polish has dried. What remains is the cast of the cuticle. The steps are as follows:

1. Place a drop of clear nail polish at one end of a clean microscope slide.

2. Place a coverslip next to the drop of nail polish, allowing the drop to run along the edge of the coverslip. Tilt the coverslip to a 45° angle, and lightly drag the coverslip across the slide to spread the polish evenly across the surface (Figure 19.7).

3. Wait a few minutes to allow the nail polish to become slightly tacky. The timing takes some practice.

4. Quickly place three or four pieces of hair onto the polish. Leave each hair hanging over the edge of the slide.

5. Allow nail polish and hair samples to dry completely for 15-20 minutes.

6. After the nail polish has dried, use forceps to grab the part of the hair that is hanging over the end of the slide, and pull the hair quickly to remove it from the slide and nail polish.

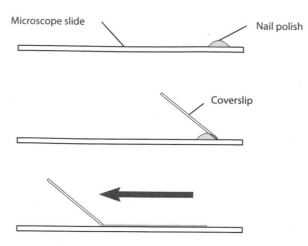

Figure 19.7 Diagram of how to create an even coating of nail polish for preparing a hair cast.

7. View the sample casts under a microscope. Begin under the lowest power and focus on a hair cast, then switch to the 10X or 40X power lens to view the details.

8. Be sure to return the lens to the smallest lens before removing your slide from the microscope stage.

9. Illustrate the scale patterns you see in your notebook, and label them with the terms from Figure 19.3 that best describe the hair cast.

For wet-mount preparations, the steps are as follows:

1. Label a clean microscope slide with the species name of the hair sample you will be working with.

2. Using a cotton swab, wet a small surface of the slide with the DPX mounting medium (or Permount).

Place the hair sample in this wet area to attach the hair to the slide. Hairs should be mounted horizontally; if hairs are small, multiple hairs can be used on the same slide (avoid overlap). Long hairs may require longer coverslips

3. Gently add several drops of mounting medium to the microscope slide (avoid moving the hair).

4. Touch one end of the cover slip onto the slide at the edge of the pool of mounting medium. Tilt the coverslip to a 45° angle (toward the hair sample), and slowly lower the coverslip down onto the mounting medium (i.e., go from 45° to 30°, then to 15°, and so on). Gently apply pressure to the other end of the coverslip (where it first touched the medium). This forces air bubbles out ahead of the coverslip and makes a more usable preparation.

5. If the mounting medium does not completely fill the cover slip, add more along the coverslip edge.

6. Allow the slide to dry completely. If you are using DPX mounting medium, drying time will be only 15 minutes or so. (If you place the slide in a drying oven, use a low temperature setting.)

7. Once the mounting medium is dry, scrape any excess off of the slide/coverslip using a single-edge razor blade (this is more difficult if you are using Permount).

8. Observe the slides under a light microscope at approximately 40X power.

9. Observe and record the medullary and cortex characteristics.

10. Estimate the medullary index (i.e., width of medulla/width of hair).

11. Take digital photographs of each hair sample for use in the analysis described in Exercise 3.

Exercise 3: Quantifying Hair Structure Using ImageJ Software

ImageJ software is introduced in Chapter 18, Exercise 3. The basic procedures are similar for hair analysis. A brief description of the analysis is given below. Before you begin the analysis, you will need digital photos of your hair samples and a digital photograph of a micrometer slide taken at the same magnification as your hair photos. You will apply image analysis techniques to your digital images, using the public domain software ImageJ, written and maintained by Wayne Rasband at the National Institute of Mental Health, Bethesda, Maryland, USA (ImageJ is the successor of NIH Image). ImageJ is written in Java, which means that it can be run on any system for which a Java runtime environment (JRE) exists (e.g., Windows, Mac, Linux). If you have not already done so, go to the ImageJ website and download the appropriate version for your operating system (http://imagej.nih.gov/ij/). Java applications will only use the memory allocated to them. Under Edit Options > Memory & Threads, you can configure the memory available

to ImageJ. The maximum memory should be set to three-quarters of the available memory on your machine.

In the first part of this exercise, you will use the photo of the micrometer slide to create a scale bar for each subsequent image. This procedure is described in Chapter 18, Exercise 3. With the scale set properly, you are ready to take measurements of your hair samples.

1. Open your first hair image (go to File > Open and browse to the correct file). Add a scale bar to the image by going to Analyze > Tools > Scale Bar. Specify the width, thickness, font size, color, and location of the scale bar, and click OK.

2. Use the magnifying tool to enlarge the region of interest if necessary. Click on the line segment tool to select it. (Find the line-segment tool by right-clicking the straight-line tool, which is fifth from left in the tool bar.)

3. For hair diameter, move the mouse cursor over the hair, click and drag from one outside edge across the hair

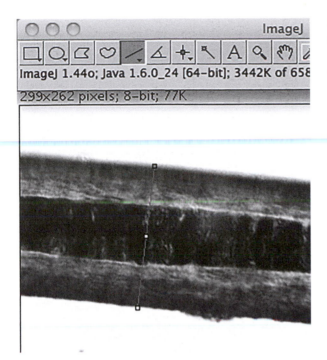

Figure 19.8 A view of the ImageJ window showing the icon for the line segment tool in the toolbar and a hair sample with a line segment drawn over it.

at 90° to the opposite outside edge, and double-click to release the line segment (Figure 19.8).

4. Create a data table of measurements for total diameter, medulla thickness, cortex thickness, etc.

5. Transfer measurements to a spreadsheet by right-clicking in the Results window, selecting Copy All, switching to the spreadsheet program, and then pasting.

Additional information about microscopic examination of hair can be found in Deedrick and Koch (2004).

Exercise 4: Extracting DNA from Hair Samples

Forensic scientists routinely extract DNA from hair samples collected at crime scenes. Wildlife biologists have adapted these techniques to answer questions about species identity, sex, and paternity from hair or feces collected in the field. Although there are a number of techniques for extracting DNA, this exercise will introduce the technique using a relatively simple kit from QIAGEN, Inc., Santa Cruz, California (Promega also makes a similar kit called DNA IQ).

Remove hair from glue plates or tape by washing in xylene until glue softens and hairs can be removed. Then follow the procedures described in the kit; a basic outline is given here:

1. Cut 15 hairs into small pieces, place in a 1.5-milliliter (ml) microcentrifuge tube, and add 200 microliters (μl) of Buffer X1. Incubate at 55°C for at least 1 hour until the sample is dissolved. Invert the tube occasionally to disperse the sample, or place on a rocking platform.

2. Add 200 μl Buffer AL and 200 μl ethanol to the sample, and mix by pulse-vortexing for 15 seconds. After mixing, briefly centrifuge the 1.5-ml microcentrifuge tube to remove drops from inside the lid.

3. Carefully apply the mixture from step 2 (including the precipitate) to the QIAamp Mini spin column (in a 2-ml collection tube) without wetting the rim. Close the cap, and centrifuge at $6000 \times g$ (8000 rpm) for 1 minute. Place the QIAamp Mini spin column in a clean 2-ml collection tube (provided), and discard the tube containing the filtrate.

4. Carefully open the QIAamp Mini spin column, and add 500μl Buffer AW1 without wetting the rim. Close the cap, and centrifuge at $6000 \times g$ (8000 rpm) for 1 minute. Place the QIAamp Mini spin column in a clean 2-ml collection tube (provided), and discard the collection tube containing the filtrate.

5. Carefully open the QIAamp Mini spin column, and add 500μl Buffer AW2 without wetting the rim. Close the cap and centrifuge at full speed ($20,000 \times g$; 14,000 rpm) for 3 minutes.

6. Place the QIAamp Mini spin column in a clean 1.5-ml microcentrifuge tube (not provided), and discard the collection tube containing the filtrate. Carefully open the QIAamp Mini spin column, and add 200 μl Buffer AE or distilled water. Incubate at room temperature for 1 minute, and then centrifuge at $6000 \times g$ (8000 rpm) for 1 minute.

7. Repeat step 6 (see also Figure 19.9). (Note: Do not elute volumes of more than 200 μl into a 1.5-ml microcentrifuge tube. For long-term storage, elute DNA in Buffer AE and store at −20°C.)

Figure 19.9
Sample protocol for extraction of DNA from hair samples using QIAmp DNA Mini Kits.

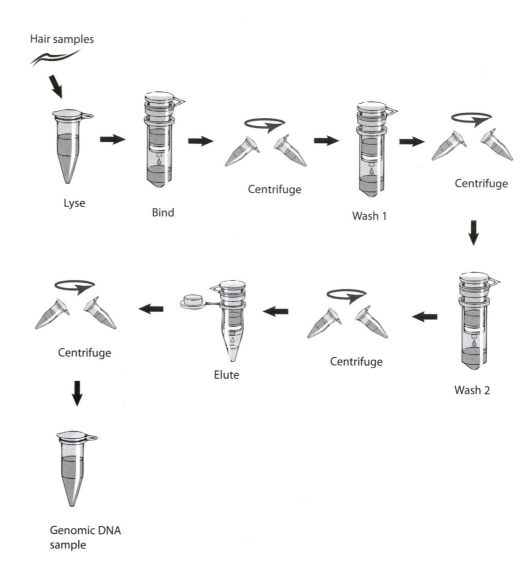

Hair samples

Lyse

Bind

Centrifuge

Wash 1

Centrifuge

Wash 2

Centrifuge

Elute

Centrifuge

Genomic DNA sample

Instructor Resources

Equipment and other resources for completing the exercises in this manual can be purchased from a variety of vendors, including those listed below. Please note that the inclusion of any manufacturer on this list should in no way be interpreted as an endorsement of their products.

General Field Equipment Sources

Forestry Suppliers, Inc.
205 West Rankin Street
PO Box 8397
Jackson, MS 39284-8397 USA
Sales: 1-800-647-5368
International: +1-601-354-3565

Ben Meadows Company
PO Box 5277
Janesville, WI 53547-5277 USA
Phone: 1-800-241-6401
Worldwide: +1-608-743-8001

Alana Ecology, Ltd.
New Street
Bishop's Castle
Shropshire SY9 5DQ
United Kingdom
Phone: +44 1588 630173
Fax: +44 1588 630176

Mammalian Skulls and Skeletons

Skulls Unlimited International
10313 S. Sunnylane Road
Oklahoma City, OK 73160 USA
1-800-659-SKULL
Sales@SkullsUnlimited.com
www.SkullsUnlimited.com

Bone Clones, Inc.
9200 Eton Ave.
Chatsworth, CA 91311
1-800-914-0091
Email: info@boneclones.com
https://boneclones.com

Biotelemetry Resources

Wildlife Telemetry Clearinghouse
http://nhsbig.inhs.uiuc.edu/wes/equipment_suppliers.html

AVM Instrument Company, Ltd.
2356 Research Drive
Livermore, CA 94550 USA
Phone: 1-510-449-2286
Fax: 1-510-449-3980
Email: avmtelem@ix.netcom.com

Advanced Telemetry Systems, Inc.
470 1st Avenue NW
Box 398
Isanti, MN 55040 USA
Phone: 1-612-444-9267
Email: 70743.512@compuserve.com

Biotrack
Stoborough Croft, Grange Road
Wareham, Dorset
BH20 5AJ
United Kingdom
Phone: +44 929 552 992
Fax: +44 929 554 948

Custom Electronics of Urbana, Inc.
2009 Silver Court W.
Urbana, IL 61801 USA
Phone: 1-217-344-3460
Fax: 1-217-344-3460

Custom Telemetry Co.
1050 Industrial Drive
Watkinsville, GA 30677 USA
Phone: 1-706-769-4024
Fax: 1-706-769-4026

Holohil Systems, Ltd.
112 John Cavanagh Road
Carp, Ontario K0A 1L0
Canada
Phone: 1-613-839-0676
Fax: 1-613-839-0675

Lotek Engineering, Inc.
115 Pony Drive
Newmarket, Ontario L3Y 7B5
Canada
Phone: 1-905-836-6680
Fax: 1-905-836-6455
Email: telemetry@lotek.com

Mariner Radar, Ltd.
Suffolk
NR32 5DN
United Kingdom
Phone: +44 502 567195
Fax: +44 502 567195

Merlin Systems, Inc.
445W Ustick Road
Meridian, ID 83642 USA
Phone: +1-208-884-3308
Fax: +1-208-888-9528
Email: merlin@cyberhighway.net

Microwave Telemetry, Inc.
10280 Old Columbia Road
Suite 260
Columbia, MD 21046 USA
Phone: 1-410-290-8672
Fax: 1-410-290-8847
Email: Microwt@aol.com

Mini-Mitter Co., Inc.
PO Box 3386
Sunriver, OR 97707 USA
Phone: 1-503-593-8639
Fax: 1-503-593-5604
Email: rrushmmtr@aol.com

Televilt International AB
Box 53
S-711 22 Lindesberg
Sweden
Phone: +46 581 17195
Fax: +46 581 17196

Telonics
932 East Impala Avenue
Mesa, AZ 85204-66990 USA
Phone: 1-602-892-4444
Fax: 1-602-892-9139

Wildlife Materials, Inc.
Route 1, Box 427A
Carbondale, IL 62901 USA
Phone: 1-618-549-6330
Fax: 1-618-457-3340

Camera Traps

TrailCamPro.com
3620 South National Avenue
Springfield, MO 65807 USA
1-800-791-0660

Camera Traps
PO Box 2006
Hillcrest
KwaZulu-Natal
3650 South Africa
Phone: +27 83 560 0555; +27 86 505 5536
Email: chris@cameratrap.co.za
http://www.cameratrap.co.za/

Bushnell Corporation
9200 Cody Street
Overland Park, KS 66214-1734 USA
Consumers: 1-800-423-3537
Fax: 1-913-752-3550

Cuddeback
PO Box 10447
Green Bay, WI 54307-0447 USA
Phone: 1-920-347-3810
Fax: 1-920-347-3820
http://www.cuddeback.com/

Sound Recording

Wildlife Sound Recording Society
https://www.wildlife-sound.org/
(many useful ideas and excellent advice)

Raven Lite Software
http://www.birds.cornell.edu/brp/raven/RavenOverview
   .html

Audacity Software
http://audacity.sourceforge.net/

Macauley Library, Cornell Lab of Ornithology
http://macaulaylibrary.org

Saul Mineroff Electronics, Inc.
574 Meachem Avenue
Elmont, NY 11003 USA
Phone: 1-516-775-1370
Fax: 1-516-775-1371
Email: info@mineroff-nature.com
http://www.mineroff.com/

Full Compass Systems
8001 Terrace Avenue
Middleton, WI 53562 USA
Phone: 1-800-356-5844
www.fullcompass.com

Marice Stith Recording Services
732 Bowling Green Road
Cortland, NY 13045 USA
Phone: 1-607-756-0145
www.stithrecording.com

Professional Sound Services
311 W. 43rd Street, Suite 1100
New York, NY 10036 USA
Phone: 1-800-883-1033
www.pro-sound.com

zZounds Music
PO Box 479
Franklin Lakes, NJ 07417-0479 USA
http://www.zzounds.com

Ultrasound Recording Equipment (Bat Detectors)

Ultra Sound Advice
27 Merton Hall Road
Wimbledon, London
SW19 3PR
United Kingdom
Phone: +44 20 8287 4614
Email: sales@ultrasoundadvice.co.uk
http://www.ultrasoundadvice.co.uk/

Wildlife Acoustics
970 Sudbury Road
Concord, MA 01742-4939 USA
Phone: 1-888-733-0200
International: +1-978-369-5225
Fax: 1-781-207-5523
Email: support2010@wildlifeacoustics.com
http://www.wildlifeacoustics.com

Pettersson
Uppsala Science Park
Dag Hammarskjolds v. 34A
S-751 83 Uppsala
Sweden
Phone: +46 1830 3880
Email: info@batsound.com
http://www.batsound.com

Batbox Ltd
2A Chanctonfold
Horsham Road
Steyning
West Sussex
BN44 3AA
United Kingdom
Phone: +44 01903 816298
Email: emailinfo@batbox.com

Titley Scientific
Email: info@titley-scientific.com
http://www.titley.com.au

Glossary

A

Acrocentric Having the centromere positioned such that one arm of a chromosome is shorter than the other.

Ad lib sampling Recording data or sampling whenever the observer deems it appropriate or interesting.

Alarm call A vocalization produced by an animal in distress or to warn other nearby animals that a predator has been detected.

Amino acid sequence The order in which amino acids lie in a chain in a peptide or protein.

Ancestral Of, or pertaining to, or inherited from, a common ancestor.

Anesthesia A method of chemical immobilization resulting in reduced sensation and awareness.

Arboreal Living mainly within the crowns of trees.

Azimuth The direction of a celestial object from the observer, expressed as the angular distance from the north or south point of the horizon to the point at which a vertical circle passing through the object intersects the horizon.

B

Baleen plates The keratinized straining plates that form from the integument of the upper jaw in mysticete whales. They are used to filter small invertebrates from seawater.

Bat detector An electronic device for detecting or recording ultrasonic signals produced by bats during echolocation.

Bayesian inference A statistical method in which Bayes's theorem is used to update hypotheses as new information is acquired.

Biotic potential The exponential growth of a population growing in an ideal and unlimited environment.

Bone marrow A soft fatty substance, in which blood cells are formed, found in the cavities of long bones.

Brachydont Pertaining to cheek teeth with low crowns; typical of mammals with omnivorous diets.

Bunodont Pertaining to low-crowned teeth with rounded or blunt cusps, used for crushing.

C

Camera trap A non-invasive capture technique that uses a camera placed in a field location to remotely photograph wild animals.

Canine A tooth posterior to the incisors and anterior to the premolars that is usually elongated, single-rooted, and single-cusped, and which is rooted in the maxilla or dentary.

Caniniform Canine-shaped.

Carnassials The cofunctioning pair of bladelike shearing teeth of carnivorans, including the last upper premolar and the first lower molar.

Cementum The relatively soft bony material on parts of a tooth in some mammals, with a structure different from enamel or dentine.

Centromeres The region of a chromosome where microtubules of a spindle attach during cell division.

Cetacean A marine mammal with a tail fluke instead of visible hind limbs and a blowhole for breathing (includes whales, dolphins, and porpoises).

Character A feature of an organism that can be described, measured, or effectively communicated between scientists; for example, pigmentation of incisor enamel.

Cheek teeth The premolar and molar teeth in mammals.

Chorioallantoic placenta A type of placenta, found in eutherians and to a lesser extent in peramelemorph metatherians, composed of an outer chorionic layer and an inner vascularized allantois.

Choriovitelline placenta A type of placenta, often called a "yolk sac placenta," found in metatherians (except bandicoots) in which there are no villi and there is only a weak connection to the uterus.

Chromatin The DNA and histone proteins of a eukaryotic cell.

Chromosome A long thread-like chain of nucleotides and proteins found in most cells and carrying the genetic information of the organism in the form of genes.

Cingulum A ridge of enamel on the base, margin, or crown of a tooth.

Clade A group of species or higher taxa consisting of a single common ancestor and all its descendants.

Cladist A person who practices the cladistic approach to phylogenetic reconstruction.

Cladistics A method of reconstructing a phylogenetic hypothesis that is based on grouping taxa solely by their shared derived character states.

Cladogram A branching diagram that illustrates hypothetical relationships between taxa and shows the evolution of lineages of organisms that have diverged from a common ancestor.

Closed population A population in which neither immigration nor emigration occurs.

Compass bearing A direction toward which you are heading as shown on a compass (or toward which the strongest radio signal is coming from in radio tracking).

Compass declination *See* Declination

Condyle A smooth rounded projection on a bone used for articulation with another bone.

Cortex The outer layer or portion of an organ, bone, or hair.

Crepuscular Active mostly near dawn and dusk.

Crest A narrow prominent ridge on a bone.

Crista A ridge formed by the fusion of cusps on a tooth.

Crustaceans A diverse group of aquatic arthropods having a hard outer shell, segmented body, and jointed limbs (crabs, lobsters, shrimp).

Cusp A prominent projection on the occlusal surface of a tooth.

D

Deciduous teeth Teeth that are replaced usually early in a mammal's life.

Declination The angle in the horizontal plane between magnetic north and true north.

Dental formula A shorthand expression of the characteristic number of each type of teeth on one side of the skull in mammals. For example, the dental formula for a soricid shrew is 3/1, 1/1, 3/1, 3/3 = 32, which indicates that this shrew has three upper incisors over one lower incisor, one upper and one lower canine, three upper over one lower premolar, and three upper and lower molars on each side of the jaw, for a total of 32 teeth on both sides.

Dentary The single bone making up one-half of the mandible, or lower jaw, of mammals.

Dentine A bonelike material that forms the body of a tooth. It is hard and made of hydroxyapatite crystals and collagen. It differs from bone in that it lacks osteocytes and osteons. The dentine of a tooth is often covered by harder enamel.

Derived Refers to a character state that is a modified version of, and differs from, that in the ancestral stock.

Diastema A space between the teeth, usually the incisors and cheek teeth. Typical of rodents and lagomorphs but also found in artiodactyls, perissodactyls, and other mammals.

Dichotomous key A set of instructions where a series of choices (couplets) between alternative characters leads progressively to the identification of a species.

Dilambdodont A pattern on the crown of an upper molar where there is a prominent W-shaped ectoloph.

Diphyodont A pattern of tooth replacement involving only two sets of teeth, typically a set of deciduous ("milk") teeth and a set of permanent teeth.

Diploid number The total number of chromosomes in the cell nucleus of a somatic cell.

Diurnal Active primarily during daylight hours.

DNA duplication Mechanism whereby new genetic information is generated through the copying of a region of DNA that contains a gene (also chromosomal duplication).

DNA sample The collection of a sample containing cells that contain DNA.

DNA sequence The precise order of nucleotides in a section of DNA.

Dominance hierarchy A type of social structure in which each animal in a group holds a rank, with some individuals dominant over others in the group and others submissive to those above them in the hierarchy.

E

Echolocation The process of emitting sounds and using the information from the returning echoes to sense the surrounding environment.

Ecotoloph A ridge connecting the paracone and metacone on a lophodont tooth.

Ectoparasite A parasite that lives on the outside of its host's body.

Enamel Hard crystalline material in the teeth of vertebrates, similar in composition to bone but without bone-forming osteocytes.

Encounter rate In optimal foraging theory, the rate at which suitable prey items are encountered by a predator; the density of suitable prey.

Endoparasite A parasite that lives on the inside of its host's body.

Ethogram A catalog of all unique behaviors observed for an animal.

Euchromatin Chromosomal material that does not stain darkly. It comprises the most active portion of the genome.

Ever-growing teeth Teeth that continue to grow throughout the life of the animal; sometimes called rootless or hypselodont teeth.

F

Field notes Any set of organized notes taken while conducting field research.

Fixative A chemical used to preserve or stabilize biological material.

Focal animal sampling In behavioral studies, when an observer chooses a single individual and records its behaviors for a period of time while ignoring behaviors of other individuals.

Folivore An animal whose primary diet consists of leaves.

Foot-drumming A rapid pounding of the foot onto the substrate to generate a seismic signal.

Foramen An opening or passage through a bone. Typically, the opening transmits nerves or blood vessels.

Fossa A shallow depression or trench on the surface of a bone.

Fossorial Pertaining to an animal that digs burrows for shelter and forages underground.

Fundamental frequency In acoustics, the lowest or root tone of a chord.

G

Genealogy A line of descent traced continuously from an ancestor.

Gnawing Biting or chewing using primarily the incisor teeth.

GPS coordinates A set of latitude and longitude coordinates generated using the Global Positioning System of satellites.

GPS tracking A technique that uses the Global Positioning System satellites to determine the precise location of an animal.

H

Handling time The amount of time from when an animal first locates a food item until it has eaten the food item.

Harmonics In acoustics, frequencies that are integral multiples of the fundamental frequency.

Hemimetabolous In insects, an incomplete metamorphosis that includes three distinct stages—egg, nymph, and adult.

Hemoparasitic A parasite with at least part of its life cycle spent in the blood.

Herbivore An animal that eats plants.

Heterochromatin Tightly packed DNA in a nucleus where gene activity is suppressed.

Heterodont Pertaining to teeth that vary in structure in different parts of the tooth row; for example, the teeth of a mammal are usually differentiated into incisors, canines, premolars, and molars.

Hibernaculum The place in which an animal hibernates.

Hibernation A period of inactivity normally induced by cold ambient temperatures characterized by lowered body temperatures and a depressed metabolic rate.

Holometabolous In insects, a complete metamorphosis including four stages—egg, larva, pupa, and adult.

Home range The area in which an individual or group spends the bulk of its time.

Homodont Pertaining to teeth that do not vary in structure in different parts of the tooth row, such as the numerous and identical conical teeth of many dolphins.

Hybridization The act or process of mating between different species to create a hybrid organism.

Hypercarnivorous Having characteristics adapted for a diet solely of meat.

Hypocone A main cusp found on the distal lingual side of the upper molars.

Hypoconid The main cusp on the labial side of the talonid of the lower molars.

Hypoconulid A prominent cusp on the posterior portion of the talonid of the lower molars.

Hypsodont Pertaining to cheek teeth with high crowns, adapted to a diet of relatively abrasive plant material.

Hystricognathous A condition in certain rodents in which the angular process of the mandible is usually lateral to the plane of the alveolus of the lower incisor.

Hystricomorphous A condition in certain rodents in which the infraorbital foramen is greatly enlarged and transmits a portion of the medial masseter muscle.

I

Incisor A tooth in mammals located anterior to the canines and rooted in the premaxilla (upper) or dentary (lower).

Infrasound Sound of very low frequency (less than 20 Hz; below the range of human hearing), such as that produced by elephants or giraffes for long-distance communication.

Insectivorous Those animals that eat primarily insects or other invertebrates.

K

Karyogram A photo or diagram depicting the chromosomes of a cell arranged in homologous pairs and numbered in sequence.

Karyotype The characteristic chromosome complement of a cell, individual, or species.

Keratin A fibrous protein forming the structure of hair, hooves, claws, horns, and baleen.

L

Labial Of or pertaining to the lips; sometimes used to refer to a structure that is on the side of the mouth adjacent to the lips.

Lactation A unique feature of mammals in which milk is formed and secreted by the female's mammary glands for nourishing the developing young after birth.

Latitude An angular distance north or south of the equator expressed in degrees and minutes.

Line On a bone, refers to a narrow raised ridge.

Linear regression Statistical method for modeling the relationship between a scalar dependent variable (y) and an independent variable (x).

Lingual Of or pertaining to the tongue; sometimes used to refer to a structure that is on the tongue side of the tooth row.

Livetrapping A method of capturing animals alive.

Longitude An angular distance east or west of the meridian at Greenwich, England, expressed in degrees and minutes.

Loph A transverse ridge on the occlusal surface of a tooth.

Lophodont Pertaining to cheek teeth in which there are a series of transverse ridges, or lophs, on the occlusal (chewing) surface.

Loxodont Tooth pattern where many parallel ridges form a washboard-like appearance.

Lunge feeding A method of feeding underwater in which the predator moves forward with its mouth open, engulfing prey and the surrounding water.

M

Mammary gland Milk-producing gland found in female mammals. The growth and activity of mammary glands are governed by several reproductive hormones.

Mandible The lower jaw.

Mark recapture A method used to estimate animal population size where a proportion of the individuals are captured, marked, released, and subsequently recaptured for counting.

Marsupial mammal The viviparous, nonplacental mammals belonging to the order Marsupialia.

Meatus A small tubular opening in a bone or group of bones.

Medulla The inner layer or core of an organ or hair.

Melanin A dark brown pigment in hair or skin.

Metacone A main cusp on the posterior, labial side of the upper molars.

Metaconid A main cusp on the posterior, lingual side of the trigonid in the lower cheek teeth.

Metaphase The stage in cell division during which the chromosomes are condensed, attached to the spindle fibers, and lined up along the equator of the cell.

Metatherian A mammal group comprising the marsupials.

Minimum convex polygon A method to enclose a set of location points such that all data points are enclosed by connecting the outer locations into a convex polygon.

Mitotic inhibitor A chemical that can arrest the process of mitosis.

Molar A cheek tooth located posterior to the premolars and having no deciduous precursor.

Molariform Pertaining to teeth that have the shape and appearance of molars but that are not molars.

Molecular characters Hereditary molecular differences, mainly in DNA or protein sequences, that are used to understand an organism's evolutionary relationships.

Molecular phylogeny A hypothetical representation of the evolutionary history of a group of organisms based on characters defined at the molecular level.

Monophyletic Refers to a group of organisms whose members are all descended from (and including) a common ancestor.

Monophyodonty Having only one generation of teeth.

Multiparous A female with embryos and one or more sets of placental scars.

Myomorphous A pattern of jaw musculature in rodents in which a slip of the medial masseter passes through an oval or V-shaped infraorbital foramen.

Mysticeti One of two suborders of cetaceans whose living members are characterized by the presence of baleen plates used for filtering food from water; baleen whales.

N

Natural selection A process whereby those organisms better adapted to their environment tend to survive and produce more offspring than those organisms that are less well adapted. This process was proposed by Charles Darwin and Alfred Russel Wallace, and it is believed to be the main process resulting in evolution.

Nocturnal Active mainly during the night.

Node Refers to the place on a cladogram where two lineages diverge.

Non-invasive sampling An approach to data collection where animals are not handled or harmed.

Nulliparous A female with no embryos or any evidence of previous placental scars.

O

Occlusal surface The surface of the tooth or tooth row that contacts the opposing surface of tooth from the opposing jaw.

Odontoblast A cell type found in the tooth cavity that produces dentine, the substance just below the tooth enamel.

Odontoceti One of two suborders of cetaceans whose living members are characterized by having teeth instead of plates of baleen; dolphins and toothed whales.

Omnivorous Having a diet consisting of both animal and plant material.

Open population A population where individuals immigrate into the population or emigrate away from the population.

Open source A software product or software code that is made freely available and where users may modify and redistribute the code for free.

Opposable thumb A thumb that can be placed opposite the fingers of the same hand allowing the digits to grasp objects.

Optimal foraging A theory in ecology that predicts that organisms will behave in such a way as to maximize their energy intake per unit time.

Oscillogram *See* Sonogram

Outgroup In phylogeny reconstruction, a group used for comparison that is related to but not part of the group under study.

P

Paracone A main cusp anterior to the protocone and on the labial side of the upper molars.

Paraconid A main cusp of the lower molars on the anterior and lingual part of the trigonid.

Parsimony Economy of explanation where the simplest explanation is considered the most likely.

Perforate Having an opening.

Phylogenetic relationship *See* Phylogeny

Phylogenetic tree *See* Phylogeny

Phylogeny The evolutionary history of an organism or group of organisms with respect to ancestor-descendant relationships; also, a hypothesis graphically describing such relationships, usually in a tree-like or bush-like diagram.

Pinniped Marine mammals that have front and rear flippers (includes seals, sea lions, walruses).

Piscivore An animal that eats primarily fish.

Placental mammals The mammalian group whose members possess a placenta. Placentals include all living mammals except marsupials and monotemes.

Polymorphism The occurrence of different forms among individuals in a population, or the presence of genetic variation within a population.

Polyphyodonty Having more than two generations of teeth.

Premolar Cheek tooth located anterior to the molars and posterior to the canines.

Primiparous A female with one set of embryos or placental scars from a recent birth.

Process On a bone, refers to a small projection or bump.

Profitability In optimal foraging theory, the amount of energy (calories) in a prey item divided by the handling time (minutes) for that food item.

Protocone The major cusp on the lingual side of upper cheek teeth.

Protoconid A main cusp on the labial side of lower cheek teeth, located at the apex of the trigonid in the molars.

Protrogomorphous The ancestral condition in some rodents in which the masseter muscles arise solely from the zygomatic arch and do not penetrate the infraorbital foramen.

Pseudo-replication The result of treating data as independent in statistical tests when the data are not independent.

Pulp cavity The space inside a tooth containing the dental pulp, which consists of living connective tissue and odontoblasts (cells that produce dentine).

Q

Quadrate molar A molar that has a square crown.

Quadritubercular An upper tooth with four main cusps—protocone, paracone, metacone, and hypocone.

R

Radio telemetry *See* Radio tracking

Radio tracking A technique used in field studies in which an animal is fitted with a battery-powered radio of a specific frequency and the observer uses a radio receiver tuned to that frequency to locate the animal.

Recent common ancestor In phylogeny, the most recent common ancestor of a group of organisms is the individual from which all other members of the group directly descended.

Recrudescence The movement of the testes into the abdomen from the scrotum, usually on a seasonal schedule.

Reproductive potential *See* Biotic potential

Rostrum A snout or anterior prolongation of the head.

S

Satellite tracking *See* GPS tracking

Scan sampling In behavioral studies, when an observer scans the entire group of animals and records what each animal is doing.

Secodont Having teeth that are blade-like.

Sectorial Pertaining to a tooth adapted for cutting.

Seismic signal Communication signal made by kangaroo rats and other burrowing rodents that involves striking the ground to produce a series of low-frequency vibrations that travel through the ground.

Selenodont An occlusal pattern of a tooth in which longitudinally arranged, crescent-shaped ridges or lophs are formed on the tooth surface; typically occurs in artiodactyls.

Semifossorial Refers to mammals that are partially, but not completely, adapted for life underground or for digging.

Septum A bony fence that separates two regions.

Sonogram A graph of a sound showing the distribution of energy at different frequencies.

Species richness The number of different species in a defined region.

Spectrogram A graph of a sound displaying frequency versus amplitude in decibels.

Stereoscopic vision The ability of an animal with two eyes to perceive a single three-dimensional image.

Study skin A specimen that has been prepared for a museum collection by removing most of the body's muscle and skeleton, replacing them with cotton, and attaching a data tag.

Stylar shelf A small cingulum or shelf that runs along the labial side of the tooth.

Submetacentric Having the centromere of a chromosome positioned such that one arm of the chromosome is slightly shorter than the other.

Sulcus A groove on a bone.

Suture In osteology, a contact line or tight joint between two bones such as the bones of the skull.

Symphysis The cartilaginous junction or articulation formed between two bones; this junction or articulation may fuse two bones together.

Symplesiomorphy In phylogenetics, an ancestral character shared by two or more groups.

Synapomorphy In phylogenetics, a derived, homologous character shared by two or more groups.

T

Talonid The "heel" or back half of a tribosphenic lower molar that occludes with the protocone of an upper molar.

Taxonomy The practice of naming and classifying organisms.

Tetrapod A four-footed vertebrate (though some tetrapods, such as whales and snakes, have lost one or both pairs of limbs).

Therians The major group of mammals that comprises marsupials and placentals.

Time budget A chart or table of the time spent doing each observed behavior.

Topographic map A map that represents relief using contour lines.

Trait A characteristic of an organism.

Transition matrix A table that summarizes the number of times (frequency) a particular behavior follows another behavior in a temporal sequence of behaviors.

Translocation A chromosomal abnormality where parts of non-homologous chromosomes are rearranged to form new combinations of genes.

Trap-nights A count of traps open for 24 hours multiplied by the number of 24-hour periods in the session.

Triangulation A method of locating radio-collared animals where the animal's position is assumed to be at the intersection of two or more compass bearings that reflect the directions of the strongest radio signal.

Tribosphenic Pertaining to a type of molar in which the protocone of a three-cusped upper tooth fits into a basin in the talonid of a lower tooth (for crushing food) and in which crests between other cusps shear past one another (for cutting food).

Trigonid The front half of a tribosphenic lower molar, which includes a triangle formed by three cusps.

Trochanter A large rounded projection of bone for muscle attachment.

U

Ultrasonic Referring to sounds at frequencies beyond human hearing capability (i.e., above 20 kilohertz).

V

Vector Any organism that transmits a pathogen.

Vigilance behavior Behavior in which an animal spends time searching its surroundings for predators or other forms of danger.

Viviparity (viviparous) The ability to give birth to live young rather than laying eggs.

Voucher specimen Any specimen that is retained as a reference and typically placed in a museum for long-term storage and use by others.

Z

Zalambdodont A type of upper molar characterized by a protocone and a V-shaped crest (an ectoloph) with the largest cusp at the apex of the V.

Zygomatic arch An arch of bone on the side of a mammal skull that is formed by the jugal bone and a process of the squamosal bone; cheekbone.

Bibliography

Agresti, A. (1994) Simple capture-recapture models permitting unequal catchability and variable sampling effort. Biometrics, 50:494-500.

Ahlén, I., & H. Baagøe. (1999) Use of ultrasound detectors for bat studies in Europe: Experiences from field identification, surveys and monitoring. Acta Chiropterologica, 1:137-150.

Altmann, J. (1974) Observational study of behavior: Sampling methods. Behaviour, 49:227-267.

Ancrenaz, M., A. J. Hearn, J. Ross, R. Sollmann, & A. Wilting. (2012) Handbook for wildlife monitoring using camera-traps. BBEC II Secretariat, Ministry of Natural Resources, Sabah, Malaysia.

Anderson, D. R., J. L. Laake, B. R. Crain, & K. P. Burnham. (1979) Guidelines for line transect sampling of biological populations. Journal of Wildlife Management, 43:70-78.

Archie, E. A., T. A. Morrison, C. A. H. Foley, C. J. Moss, & S. C. Alberts. (2006) Dominance rank relationships among wild female African elephants, *Loxodonta africana*. Animal Behaviour, 71:117-127.

Baillargeon, S., & L.-P. Rivest. (2007) Rcapture: Loglinear models for capture-recapture. Journal of Statistical Software, 19:1-31.

Balme, G. A., L. T. B. Hunter, & R. Slotow. (2009) Evaluating methods for counting cryptic carnivores. Journal of Wildlife Management, 73:433-441.

Barnes, R. F. W. (2001) How reliable are dung counts for estimating elephant numbers? African Journal of Ecology, 39:1-9.

Barnes, R. F. W., & K. L. Jensen. (1987) How to count elephants in forests. IUCN African Elephant and Rhino Specialist Group, Causeway, Zimbabwe.

Bartos, L. (1986) Dominance and aggression in various sized groups of red deer stags. Aggressive Behavior, 12:175-182.

Batcheler, C. L. (1975) Development of a distance method for deer census from pellet groups. Journal of Wildlife Management, 39:641-652.

Belovsky, G. E., & J. B. Slade. (1986) Time budgets of grassland herbivores: Body size similarities. Oecologia, 70:53-62.

Berg, S. S. (2015) The package "sigloc" for the R software: A tool for triangulating transmitter locations in ground-based telemetry studies of wildlife populations. Bulletin of the Ecological Society of America, 96:500-507. doi:10.1890/0012-9623-96.3.500.

Berg, W., A. Johnels, B. Sjostrand, & T. Westermark. (1966) Mercury content in feathers of Swedish birds from the past 100 years. Oikos, 17:71-83.

Boyd, R., & J. B. Silk. (1983) A method for assigning cardinal dominance ranks. Animal Behaviour, 31:45-58.

Braun, W., & D. Murdoch. (2007) A first course in statistical programming with R. Cambridge University Press, Cambridge, UK.

Buckland, S. T., D. R. Anderson, K. P. Burnham, & J. L. Laake. (1993) Distance sampling: Estimating abundance of biological populations. London: Chapman and Hall.

Calenge, C. (2006) The package adehabitat for the R software: A tool for the analysis of space and habitat use by animals. Ecological Modelling, 197:516-519.

Carnes, J. (2007) UTM using your GPS with the Universal Transverse Mercator coordinate system. San Carlos, CA: MapTools.

Charnov, E. L. (1976) Optimal foraging: Attack strategy of a mantid. American Naturalist, 110:141-151.

Coleing, A. (2009) The application of social networking theory to animal behavior. Bioscience Horizons, 2:32-43.

Cormack, R. M. (1985) Examples of the use of GLIM to analyse capture-recapture data. In Statistics in ornithology, B. J. T. Morgan and P. M. North, eds., 243-273. Springer-Verlag, New York.

Craighead, F. C. (1982) Track of the grizzly. Random House, New York.

Craighead, J. J., J. S. Sumner, & J. A. Mitchell. (1995) The grizzly bears of Yellowstone: Their ecology in the Yellowstone ecosystem. Island Press, Washington, DC.

Cross, P. C., J. A. Bowers, C. T. Hay, J. Wolhuter, P. Buss, J. Hofmeyr, J. T. du Toit, & W. M. Getz. (2016) Data from: Nonparameteric kernel methods for constructing home ranges and utilization distributions. Movebank Data Repository. doi:10.5441/001/1.j900f88t.

Csardi, G., & T. Nepusz. (2006) The igraph software package for complex network research. InterJournal of Complex Systems 1695. http://igraph.org.

Dalgaard, P. (2008) Introductory statistics with R. 2nd ed. Springer, New York.

Dawson, T. J. (1995) Kangaroos: Biology of the largest marsupials. Comstock Publishing Associates, Cornell University Press, Ithaca, NY.

Deedrick, D., & S. Koch. (2004) Microscopy of hair part II: A practical guide and manual for animal hairs. Forensic Science Communications, 6 (3). https://archives.fbi.gov/archives/about-us/lab/forensic-science-communications/fsc/july2004/research/2004_03_research02.htm.

Dick, C. W., & D. Gettinger. (2005) A faunal survey of streblid bat flies (Diptera: Streblidae) associated with bats in Paraguay. Journal of Parasitology, 91:1015-1024.

Dick, C. W., & B. D. Patterson. (2006) Bat flies: Obligate ectoparasites of bats. In Micromammals and macro-parasites: From evolutionary ecology to management, S. Morand, B. R. Krasnov, & R. Poulin, eds. Springer-Verlag, Tokyo.

Dittmar, K., M. L. Porter, S. Murray, & M. F. Whiting. (2006) Molecular phylogenetic analysis of nycteribiid and streblid bat flies (Diptera: Brachycera, Calyptratae): Implications for host associations and phylogeographic origins. Molecular Phylogenetics and Evolution, 38:155-170.

Domning, D. P., & L. A. C. Hayek. (1984) Horizontal tooth replacement in the Amazonian manatee (*Trichechus inunguis*). Mammalia, 48:105-128.

Faraway, J. J. (2005) Linear models with R. Chapman and Hall/CRC, Boca Raton, FL.

Farine, D. R., & H. Whitehead. (2015) Constructing, conducting and interpreting animal social network analysis. Journal of Animal Ecology, 84:1144-1163.

Feldhamer, G. A., L. C. Drickamer, S. H. Vessey, J. F. Merritt, & C. Krajewski. (2015) Mammalogy: Adaptation, diversity, ecology. 4th ed. Johns Hopkins University Press, Baltimore.

Ferguson-Smith, M. A., & V. Trifonov. (2007) Mammalian karyotype evolution. Nature, 8:950-962.

Ferreras, P., F. Diaz-Ruiz, P. C. Alves, & P. Monterroso. (2017) Optimizing camera-trapping protocols for characterizing mesocarnivore communities in south-western Europe. Journal of Zoology, 301:23-31.

Gardener, M. (2012) Statistics for ecologists using R and Excel: Data collection, exploration, analysis, and presentation. Pelagic Publishing, Exeter, UK.

Gardener, M. (2014) Community ecology: Analytical methods using R and Excel. Pelagic Publishing, Exeter, UK.

Getz, W. M., S. Fortmann-Roe, P. C. Cross, A. J. Lyons, S. J. Ryan, & C. C. Wilmers. (2007) LoCoH: Nonparameteric kernel methods for constructing home ranges and utilization distributions. PLoS ONE 2(2):e207.

Glass, B. P., & M. L. Thies. (1997) A key to the skulls of North American mammals. 3rd ed. Self-published.

Goodall, J. (1986) Chimpanzees of Gombe. Harvard University Press, Cambridge, MA.

Greene, E., & T. Meagher. (1998) Red squirrels, *Tamiasciurus hudsonicus*, produce predator-class specific alarm calls. Animal Behaviour, 55:511-518.

Grutzner, F., W. Rens, E. Tsend-Ayush, N. El-Mogharbel, P. C. M. O'Brien, R. C. Jones, M. A. Ferguson-Smith, & J. A. Marshall Graves. (2004) In the platypus a meiotic chain of ten sex chromosomes share genes with the bird Z and mammal X chromosomes. Nature, 432:913-917.

Gurnell, J., & J. R. Flowerdew. (2006) Live trapping small mammals: A practical guide. Mammal Society Occasional Publications 3. Mammal Society, London.

Hafner, M. S. (2007) Field research in mammalogy: An enterprise in peril. Journal of Mammalogy, 88:1119-1128.

Hafner, M. S., & S. A. Nadler. (1988) Phylogenetic trees support the coevolution of parasites and their hosts. Nature, 332:258-259.

Hall, B. G. (2006) Phylogenetic trees made easy: A how-to manual. 3rd ed. Sinauer Associates, Sunderland, MA.

Hall, B. K., & B. Hallgrimsson. (2008) Strickberger's evolution: The integration of genes, organisms and populations. 4th ed. Jones & Bartlett Publishing, Sudbury, MA.

Hamel, S., S. T. Killengreen, J. A. Henden, N. E. Eide, L. Roed-Eriksen, R. A. Ims, & N. G. Yoccoz. (2013) Towards good practice guidance in using camera-traps in ecology: Influence of sampling design on validity of ecological inferences. Methods in Ecology and Evolution, 4:105-113.

Handcock, R. N., D. L. Swain, G. J. Bishop-Hurley, K. P. Patison, T. Wark, P. Valencia, P. Corke, & C. J. O'Neill. (2009) Monitoring animal behaviour and environmental interactions using wireless sensor networks, GPS collars and satellite remote sensing. Sensors 9:3583-3603.

Härkönen, S., & R. Heikkilä. (1999) Use of pellet group counts in determining density and habitat use of moose *Alces alces* in Finland. Wildlife Biology, 5:233-239.

Hayes, T. B., A. Collins, M. Lee, M. Mendoza, N. Noriega, A. Stuart, & A. Vonk. (2002) Hermaphroditic, demasculinized frogs after exposure to the herbicide atrazine at low ecologically relevant doses. Proceedings of the National Academy of Sciences USA, 99:5476-5480.

Hebblewhite, M., & D. T. Haydon. (2010) Distinguishing technology from biology: A critical review of the use of GPS telemetry data in ecology. Philosophical Transactions of the Royal Society B, 365:2303-2312.

Hedges, S., M. J. Tyson, A. F. Sitompul, M. F. Kinnaird, & D. A. Gunaryadi. (2005) Distribution, status, and conservation needs of Asian elephants (*Elephas maximus*) in Lampung Province, Sumatra, Indonesia. Biological Conservation, 124:35-48.

Hersh, M. H., R. S. Ostfeld, D. J. McHenry, M. Tibbetts, J. L. Brunner, M. E. Killilea, K. LoGiudice, K. A. Schmidt, & F. Keesing. (2014) Co-infection of black-legged ticks with *Babesia microti* and *Borrelia burgdorferi* is higher than expected and acquired from small mammal hosts. PLoS ONE 9(6):e99348. doi:10.1371/journal.pone.0099348.

Hinchliff, C. E., et al. (2015) Synthesis of phylogeny and taxonomy into a comprehensive tree of life. Proceedings of the National Academy of Sciences, 112:12764-12769. http://dx.doi.org/10.1073/pnas.1423041112.

Hunter, D. M., & J. M. Webster. (1974) Effects of cuterebrid larval parasitism on deer mouse metabolism. Canadian Journal of Zoology, 52:209-217.

Hurlbert, S. H. (1984) Pseudoreplication and the design of ecological field experiments. Ecological Monographs, 54:187-211.

Jameson, K. A., M. C. Appleby, & L. C. Freeman. (1999) Finding an appropriate order for a hierarchy based on probabilistic dominance. Animal Behaviour, 57:991-998.

Jansen, P. A., T. D. Forrester, & W. J. McShea. (2014) Protocol for camera-trap surveys of mammals at

CTFS-ForestGEO sites. Smithsonian Tropical Research Institute, Panama City, Panama.

Karanth, K. U. (1995) Estimating tiger (*Panthera tigris*) populations from camera-trap data using capture-recapture models. Biological Conservation, 71:333-338.

Karanth, K. U., & J. D. Nichols. (1998) Estimation of tiger densities in India using photographic captures and recaptures. Ecology, 79:2852-2862.

Karanth, K., J. Nichols, N. Kumar, & J. Hines. (2006) Assessing tiger population dynamics using photographic capture-recapture sampling. Ecology, 87:2925-2937.

Karanth, K. U., J. D. Nichols, N. S. Kumar, W. A. Link, & J. E. Hines. (2004) Tigers and their prey: Predicting carnivore densities from prey abundance. Proceedings of the National Academy of Sciences 101:4854-4858. http://www.ncbi.nlm.nih.gov/pmc/ articles/ PMC387338/.

Kays, R. W., & K. M. Slauson. (2008) Remote cameras. In Noninvasive survey methods for carnivores, R. A. Long, P. MacKay, W. J. Zielinsky, and J. C. Ray, eds., 110-140. Island Press, Washington, DC.

Kays, R. W., S. Tilak, B. Kranstauber, P. A. Jansen, C. Carbone, M. Rowcliffe, T. Fountain, J. Eggert, & Z. He. (2011) Camera traps as sensor networks for monitoring animal communities. International Journal of Research and Reviews in Wireless Sensor Networks, 1:19-29.

Kendall, K. C., and K. S. McKelvey. (2008) Hair collection. In Noninvasive survey methods for carnivores, R. A. Long, P. MacKay, W. J. Zielinsky, and J. C. Ray, eds., 141-182. Island Press, Washington, DC.

Kenward, R. (1987) Wildlife radio tagging: Equipment, field techniques and data analysis. Academic Press, New York.

Kingdon, J. (1979) East African mammals: An atlas of evolution in Africa. Volume 3, Part B, Large mammals. University of Chicago Press, Chicago.

Kjellstrom, B., & C. Kjellstrom Elgin. (2009) Be expert with map and compass: The complete orienteering handbook. John Wiley & Sons, Hoboken, NJ.

Knowles, S. C. L., A. Fenton, O. L. Petchey, T. R. Jones, B. Barber, & A. B. Pedersen. (2013) Stability of within-host parasite communities in a wild mammal system. Proceedings of the Royal Society B, 280:0130598. http://dx.doi.org/10.1098/rspb.2013.0598.

Kolmes, S., K. Mitchell, & J. Ryan. (1997) Optimal foraging theory. Journal of Undergraduate Mathematics and Its Applications, 18:43-85.

Kouakou, C. Y., C. Boesch, & H. Kuehl. (2009) Estimating chimpanzee population size with nest counts: Validating methods in Tai National Park. American Journal of Primatology, 71:447-457.

Krebs, C. J. (1999) Ecological methodology. 2nd ed. Benjamin/ Cummings, Menlo Park, CA.

Krebs, J. R. (1978) Optimal foraging: Decision rules for predators. In Behavioural ecology: An evolutionary approach, J. R. Krebs and N. B. Davies, eds. Sinauer Associates, Sunderland, MA.

Lancia, R. A., J. D. Nichols, & K. H. Pollock. (1994) Estimating the number of animals in wildlife populations. In Research and management techniques for wildlife and habitats, 215-253. 5th ed. Wildlife Society, Bethesda, MD.

Letunic, I., & P. Bork. (2016) Interactive tree of life (iTOL) v3: An online tool for the display and annotation of phylogenetic and other trees. Nucleic Acid Research, 8:W242-W245.

Long, R. A., P. MacKay, W. J. Zielinski, & J. C. Ray. (2008) Noninvasive survey methods for carnivores. Island Press, Washington, DC.

Mandujano, S. (2014) PELLET: An Excel-based procedure for estimating deer population density using the pellet-group counting method. Tropical Conservation Science, 7:308-325.

Marques, F., S. T. Buckland, D. Goffin, C. E. Dixon, D. L. Borchers, B. A. Mayle, & A. J. Peace. (2001) Estimating deer abundance from line transect surveys of dung: Sika deer in southern Scotland. Journal of Applied Ecology, 38:349-363.

McComb, K., D. Reby, L. Baker, C. Moss, & S. Sayialel. (2003) Long-distance communication of acoustic cues to social identity in African elephants. Animal Behaviour, 65:317-329.

Merritt, J. F. (2010) The biology of small mammals. Johns Hopkins University Press, Baltimore.

Moore, S. L., & K. Wilson. (2002) Parasites as a viability cost of sexual selection in natural populations of mammals. Science, 297:2015-2018.

Mooty, J. J., P. D. Karns, & D. M. Heisey. (1984) The relationship between white-tailed deer track counts and pellet-group surveys. Journal of Wildlife Management, 48:275-279.

Morand, S., & P. H. Harvey. (2000) Mammalian metabolism, longevity and parasite species richness. Proceedings of the Royal Society of London B, 267:1999-2003.

Mowat, G., & D. Paetkau. (2002) Estimating marten *Martes americana* population size using hair capture and genetic tagging. Wildlife Biology, 8:201-209.

Muenchen, R. A. (2009) R for SAS and SPSS users. Springer Series in Statistics and Computing. Springer, New York.

Murrell, P. (2005) R graphics. Chapman and Hall/CRC, Boca Raton, FL.

Nams, V. O. (1989) Effects of radio-telemetry location error on sample size and bias when testing for habitat selection. Canadian Journal of Zoology, 67:1631-1636.

National Research Council. (1996) Guide for the care and use of laboratory animals. National Academies Press, Washington, DC.

Newman, J. A., & T. Caraco. (1987) Foraging, predation hazard and patch use in grey squirrels. Animal Behaviour, 35:1804-1813.

Niedballa, J., R. Sollmann, A. Courtiol, & A. Wilting. (2016) camtrapR: An R package for efficient camera trap data management. Methods in Ecology and Evolution, 7:1457-1462. doi:10.1111/2041-210X.12600.

Norscia, I., & E. Palagi. (2015) The socio-matrix reloaded: From hierarchy to dominance profile in wild lemurs. PeerJ 3:e729. doi:10.7717/peerj.729.

O'Brien, T. G., & M. F. Kinnaird. (2011) Estimation of species richness of large vertebrates using camera traps: An example from an Indonesian rainforest. In Camera traps in animal ecology: Methods and analyses, A. F. O'Connell, J. D. Nichols, and K. U. Karanth, eds., 233-252. Springer, Tokyo.

Otto, M. L. (2014) Estimation of white-tailed deer (*Odocoileus virginianus*) population densities in Miami University's natural areas using distance sampling. MS thesis. Miami University, Oxford, Ohio.

Patterson, B. D., C. W. Dick, & K. Dittmar. (2009) Nested distributions of bat flies (Diptera: Streblidae) on Neotropical bats: Artifact and specificity in host-parasite studies. Ecography, 32:481-487.

Payne, K. B., W. R. Langbauer Jr., & E. M. Thomas. (1986) Infrasonic calls of the Asian elephant (*Elephas maximus*). Behavioral Ecology and Sociobiology, 18:297-301.

Plumptre, W. J. (2000) Monitoring mammal populations with line transect techniques in African forests. Journal of Applied Ecology, 37:356-368.

Pocock, M. J. O., & N. Jennings. (2006) Use of hair tubes to survey for shrews: New methods for identification and quantification of abundance. Mammal Review, 36:299-308.

Pollock, K. H., S. R. Winterstein, & M. J. Conroy. (1989) Estimation and analysis of survival distributions for radio-tagged animals. Biometrics, 45:99-109.

Poole, J. H., K. Payne, W. R. Langbauer Jr., & C. J. Moss. (1988) The social contexts of some very low frequency calls of African elephants. Behavioral Ecology and Sociobiology, 22:385-392.

Presley, S. J., T. Dallas, B. T. Klingbeil, & M. R. Willig. (2015) Phylogenetic signals in host-parasite associations in Neotropical bats and Nearctic desert rodents. Biological Journal of the Linnean Society, 116:312-327.

QIAamp® DNA Mini and Blood Mini Handbook. 3rd ed. http://www.qiagen.com/literature/render.aspx?id=200373.

Randall, J. A. (1989) Individual footdrumming signatures in bannertailed kangaroo rats, *Dipodomys spectabilis*. Animal Behaviour, 38:620-630.

Randall, J. A. (1997) Species-specific footdrumming in kangaroo rats: *Dipodomys ingens, D. deserti, D. spectabilis*. Animal Behaviour, 54:1167-1175.

Randall, J. A., & M. D. Matocq. (1997) Why do kangaroo rats (*Dipodomys spectabilis*) footdrum at snakes? Behavioral Ecology, 8:404-413.

R Core Team. (2013) R: A language and environment for statistical computing. R Foundation for Statistical Computing, Vienna, Austria. http://www.R-project.org/.

Reiczigel, J., L. Rozsa, A. Reiczigel, & I. Fabian. (2013) Quantitative parasitology (QPweb). http://www2.univet.hu/qpweb.

Rexstad, E., & K. P. Burnham. (1991) User's guide for interactive program CAPTURE. Colorado Cooperative Fish and Wildlife Research Unit, Colorado State University, Fort Collins.

Rocha, L. A., & 121 others. (2004) Specimen collections: An essential tool. Science, 344:814-815.

Rogers, L. L. (1987) Seasonal changes in defecation rates of free-ranging white-tailed deer. Journal of Wildlife Management, 51:330-333.

Rouco, C., S. Santoro, M. Delibes-Mateos, & R. Villafuerte. (2016) Optimization and accuracy of faecal pellet count estimates of population size: The case of European rabbits in extensive breeding nuclei. Ecological Indicators, 64:212-216.

Rowcliffe, J. M., J. Field, S. T. Turvey, & C. Carbone. (2008) Estimating animal density using camera traps without the need for individual recognition. Journal of Applied Ecology, 45:1228-1236.

Rozsa, L., J. Reiczigel, & G. Majoros. (2000) Quantifying parasites in samples of hosts. Journal of Parasitology, 86:228-232.

Sawyer, T. G., R. L. Marchinton, & W. M. Lentz. (1990) Defecation rates of female white-tailed deer in Georgia. Wildlife Society Bulletin, 18:16-18.

Seaman, D. E., B. Griffith, & R. A. Powell. (1998) KERNELHR: A program for estimating animal home ranges. Wildlife Society Bulletin, 26:95-100.

Seaman, D. E., & R. A. Powell. (1996) An evaluation of the accuracy of kernel density estimators for home range analysis. Ecology, 77:2075-2085.

Seber, G. A. F. (1982) The estimation of animal abundance and related parameters. 2nd ed. Macmillan, New York.

Shannon, G., J. S. Lewis, & B. D. Gerber. (2014) Recommended survey designs for occupancy modelling using motion-activated cameras: Insights from empirical wildlife data. PeerJ, 2:e532. doi:10.7717/peerj.532.

Sikes, R. S., & Animal Care and Use Committee of the American Society of Mammalogists. (2016) 2016 guidelines of the American Society of Mammalogists for the use of wild mammals in research and education. Journal of Mammalogy, 97:663-688.

Smith, A. D. (1964) Defecation rates of mule deer. Journal of Wildlife Management, 28:435-444.

Stevens, M. H. (2009) A primer of ecology with R. Springer-Verlag, New York.

Suarez, A. V., & N. D. Tsutsui. (2004) The value of museum collections for research and society. BioScience, 54:66-74.

Sutherland, W. J. (1996) Ecological census techniques: A handbook. Cambridge University Press, Cambridge, UK.

Theraulaz, G., J. Gervet, B. Thon, M. Pratte, & S. Semenoff-Tian-Chansky. (1992) The dynamics of colony organization in the primitively eusocial wasp *Polistes dominulus* (Christ). Ethology, 91:177-202.

Timm, R. M., & R. E. Lee Jr. (1981) Do bot flies, Cuterebra (Diptera: Cuterebridae), emasculate their hosts? Journal of Medical Entomology, 18:333-336.

Tomkiewicz, S. M., M. R. Fuller, J. G. Kie, & K. K. Bates. (2010) Global positioning system and associated technologies in animal behaviour and ecological research. Philosophical Transactions of the Royal Society B, 365:2163-2176.

UniProt Consortium. (2014) UniProt: A hub for protein information. Nucleic Acid Research, 43:D204-D212. doi:10.1093/nar/gku989.

Vaughan, T., J. M. Ryan, & N. Czaplewski. (2015) Mammalogy. 6th ed. Jones & Bartlett Publishing, Sudbury, MA.

Veyrunes, F., P. D. Waters, P. Miethke, W. Rens, D. McMillan, A. E. Alsop, F. Grützner, J. E. Deakin, C. M. Whittington, K. Schatzkamer, C. L. Kremitzki, T. Graves, M. A. Ferguson-Smith, W. Warren, & J. A. Marshall Graves. (2008) Bird-like sex chromosomes of platypus imply recent origin of mammal sex chromosomes. Genome Research, 18:965-973.

Ward, D. (2007) Modelling the potential geographic distribution of invasive ant species in New Zealand. Biological Invasions, 9:723-735.

Weilgart, L., & H. Whitehead. (1997) Group-specific dialects and geographical variation in coda repertoire in South Pacific sperm whales. Behavioral Ecology and Sociobiology, 40:277-285.

White, G. C., K. P. Burnham, D. L. Otis, & D. R. Anderson. (1978) User's manual for program CAPTURE. Utah State University Press, Logan.

Williams, B. K., J. D. Nichols, & M. J. Conroy. (2001) Analysis and management of animal populations. Academic Press, San Diego, CA.

Williams, E. S., & I. K. Barker. (2001) Infectious diseases of wild mammals. Iowa State University Press, Ames.

Wilson, D. E., & D. M. Reeder. (2005) Mammal species of the world: A taxonomic and geographic reference. 3rd ed. Johns Hopkins University Press, Baltimore.

Wittemyer, G., W. M. Getz, F. Vollrath, & I. Douglas-Hamilton. (2007) Social dominance, seasonal movements, and spatial segregation in African elephants: A contribution to conservation behavior. Behavioral Ecology and Sociobiology, 61:1919-1931.

Xingfeng, S., R. Kays, & P. Ding. (2014) How long is enough to detect terrestrial animals? Estimating the minimum trapping effort on camera traps. PeerJ, 2:e374. https://doi.org/10.7717/peerj.374.

Yates, T. L., J. N. Mills, C. A. Parmenter, T. G. Ksaizek, R. R. Parmenter, J. R. Vande Castle, C. H. Calisher, S. T. Nichol, K. D. Abbott, J. C. Young, M. L. Morrison, B. J. Beaty, J. L. Dunnum, R. J. Baker, J. Salazar-Bravo, & C. J. Peters. (2002) The ecology and evolutionary history of an emergent disease: Hantavirus pulmonary syndrome. BioScience, 52:989-998.

Index

Page numbers in italics refer to illustrations. Page numbers followed by the letter *t* refer to tables.

hydrophones, 119
hypercarnivorous, 15
hypocone, 18, *18, 20, 21*
hypoconid, *17,* 18
hypoconulid, *17,* 18
hypostome, *58, 59*
hypsodont, 19-20
hystricognathous, 9, *9*
hystricognathy, *9*
hystricomorphous, *8, 9*
Hystrix cristata, 20

IACUC, *43, 96, 98, 146*
igraph, 134-135, *135,* 136-137
Iguana iguana, 28; iguana, 27, *27,* 27t, 28, *28,* 28t, 29, 29t
ImageJ, 148-149, *149,* 150, *150,* 152, 156, *157*
immobilization, 44, 96, 98; immobilizing, 99
incisive foramina. *See* foramen
incisor, 5-6, *6,* 10, 15, *15,* 16, *17,* 19, *19,* 20-21, *21,* 22, *22*
Indiana bat, 117
infraorbital foramen. *See* foramen
infrasonic, 118-119
Insectivora, 13
insectivore, 42; insectivorous, 18-19
interorbital breadth, 10, *11*
interparietal, 3, *4*
intrauterine glands, 60
iTOL, *33*
Ixodes, 57
Ixodidae, 57

jaguar, 92
Java, 148-149, 156; applet, 132, *133*
Jolly-Seber, 62, 64-65, 68, 68t, 69t, 71, 72t, 77-78
jugal, 3, *4, 6, 9, 17*

kangaroo, 22, 27, *27,* 27t, 28, *28,* 28t, 29t; eastern gray, 28; rats, 6, 118, 138
karyogram, 145
karyotype, 145, 146, *146,* 148, 150
keratin, 20, 152
kernel estimates, 97

labial, 3, 16, 18, *18, 20*
lacrimal, 3, *4, 5, 6, 9, 17, 21*
lactation, 46
Landau's *H,* 132
latitude, 35, *36,* 38, *38,* 39, 91, 106, 108-109, 111, *112,* 114
lemurs, 16, 134, 135, 135t, 136, *136*
length of mandibular tooth row, 10, 23t
length of maxillary tooth row, 10, *11,* 23t
leopard, 92, 124; cat, 93-94, *94,* 95
Lepus americanus, 77
Levan, 149-150, *150*

lice, 55-56, 58
Lincoln-Petersen method, 62-64, 66, 71t, 74
linear regression, 143
line transects. *See* transect
lingual, 3, 16, *17,* 18, 20, *20*
lion, 92, 120, *120,* 121, *121,* 123, *123,* 124, *124,* 125
little brown bat, *17, 18*
livetrapping, 41-43, 47t, 51, 63, 76, 88; grid, 43
longitude, 35, *36,* 38-39, 91, 106, 108-109, 111, *112,* 114
lophodont, 20
lophs, 17-20
louse, *59*
loxodont, 20, *21*
Loxodonta africana, 22
lunge feeding, 21
Lyme disease, 57

Macropus giganteus, 28
Madagascar, 34-35, *36,* 42
Mammalia, 15, 32, *32,* 152
mammary glands, 27, *27,* 27t, *28,* 50
manatees, 20, 22
mandible, 3, 5, 9, 10; length, 10, *10,* 23t
marginal value theorem, 141
mark-recapture, 41-44, 62-63, 64, 64t, 65t, 66-68, 68t, 73-74, 74t, 75, 77, 91, 93
marsupials, 18
marsupium, 27t, *28*
Martes americana, 154
masked shrew, 46t, 140, 140t
masseter, 5-6, *7, 8, 9*
maxilla, 3, *4, 5, 5, 6, 9,* 10, 16, *17, 21,* 23
maxillary tooth row, 10, *11, 23, 23t*
maximum likelihood, 25, 73-74, 105
meadow vole, 46t, 65t, 140t
meatus, *4, 5, 6, 21*
medulla, 153, *153,* 156-157
Megaderma lyra, 28
Megaptera novaeangliae, 21
melanin, 153
meridian, 38-39
mesostyle, *18, 20*
metacentric, 146, 148, 150
metacone, *17,* 18, *18, 20, 21*
metaconid, *17, 18, 18*
metaconule, *17*
metastyle, *18*
metatherians, 22
Microcebus rufus, 42
micrometer, 148-149, 156
microphone, 118-120, 123
Microtus, 66t, 68, 68t, 69t, 70t, 71t; *pennsylvanicus,* 49, 46t, 65t
milk teeth, 21
minimum convex polygon, 104, *104,* 106, *106, 107*
mink, 153

mites, 57-58, *59*
mitotic inhibitor, 145-147
molar, 10, 15-16, *16,* 17, *17,* 18, *18,* 19-21, *21,* 22, *22,* 23
molariform, 20
molecular characters, 1, 9, 25
moles, 18, 41
monophyletic clades, 25
monophyodonty, 13
Monotremata, 19; monotremes, 19, 145
moose, 20, 62, 82, 138
Morganucodon, 15
mountain: beaver, 6; lion, 92
mouse lemur, *42*
MoveBank, 108, 115
multiparous, 50
museum, collection, 37, 48; specimens, 48-49, 51
muskrat, *19*
Myodes gapperi, 49
myomorphous, *8, 9*
Myotis, daubentonii, 124, *125; lucifugus, 17, 18*
Myrmecophagidae, 19
Mysticete, 20

narwhals, 16
nasal, 3, *4, 5, 5, 6, 9, 9,* 10, *17, 21, 52,* 53
natural selection, 3, 138
nematodes, 57
net caloric intake rate, 139-141, *141*
net caloric value, 138, 139
net energy, 138, 140, 141, *141*
Newick, 32-33
nocturnal, 15, 41, 44, 62, 81, *92,* 98
node, 28, *28,* 32-33, 134, 135, *135,* 136, *136,* 137
non-invasive, 37, 45, 88, 152
northing, 39, 105
nulliparous, 50
Nyctalus, 122
Nycteribiidae, 60
nymphs, 58, *58*

occipital, 3, *4, 5, 6,* 10, *21;* condyle, *4, 5, 6, 9, 17, 21*
occipitonasal length, 10, *10*
occlusal, *16, 17, 18, 19, 20, 21;* surface, 15-17, 19
occupancy, 89, 152
ocelot, 92, *92*
Odocoileus, hemionus, 130t; *virginianus,* 130t
Odontocetes, 9, 15, 19, 22
Oestridae, 59
omnivorous, 19
Ondatra zibethicus, 19
open population, 62, 64, 77-78
open source, 2, 120, 152
Open Tree of Life, 32, *32, 33*
opossum, 18, *92*
opposable thumbs, 27t, *28*

About the Author

James M. Ryan is a mammalogist. He is a coauthor of *Mammalogy*, a college textbook (sixth edition, 2015) published by Jones & Bartlett Publishers. He is also the author of *Adirondack Wildlife: A Field Guide*, published by the University Press of New England. In addition, Ryan has published numerous scientific articles in professional journals. He has conducted field research around the world, including at sites in Ecuador, Ghana, Kenya, Madagascar, Trinidad, and Uganda. His research has been funded by the National Science Foundation, the National Geographic Society, and other funding agencies.

Ryan is currently a professor of biology at Hobart and William Smith Colleges in Geneva, New York. He received his PhD in zoology from the University of Massachusetts, a master's degree in biological sciences from the University of Michigan, and a bachelor's degree in zoology from the State University of New York at Oswego.

Additional information about the author and about mammals can be found at his blog *Wild Mammal* at http://www.wildmammal.com.